country, native title and ecology

country, native title and ecology

Edited by Jessica K Weir

Australian
National
University

E PRESS

Published by ANU E Press and Aboriginal History Incorporated
Aboriginal History Monograph 24

This title is also available online at: http://epress.anu.edu.au/

National Library of Australia Cataloguing-in-Publication entry

Title: Country, native title and ecology / edited by Jessica K Weir.

ISBN: 9781921862557 (pbk.) 9781921862564 (ebook)

Series: Aboriginal history monograph ; 24

Subjects: Native title (Australia)--Economic aspects.
                Environmental management
                Aboriginal Australians--Ethnic identity.
                Aboriginal Australians--Land tenure.
                Aboriginal Australians--Environmental aspects.

Other Authors/Contributors:
                Weir, Jessica K.

Dewey Number:305.89915

Cover design and layout by ANU E Press

Front Cover photo: Karajarri Head Ranger John Hopiga directing at Gourdon Bay, West Kimberley, photo by Jessica Weir.

Back Cover photos: Ivan Namirrkki with his family, photo by Luke Taylor; Banula Marika and Ben Hoffmann, photo by Ben Hoffmann; and Pampila Boxer and Warford Bujiman with a rain-making stone, at Yakanarra, West Kimberley, photo by Patrick Sullivan.

Contents

List of Figures and Tables

List of Figures

List of Tables

Acknowledgements

Country, Native Title and Ecology comprises of chapters directly solicited from authors, as well as papers that developed out of presentations from two AIATSIS conference sessions I convened: 'Native title and environmental change', *National Native Title Conference: Tides of Native Title*, 6-8 June 2007, Cairns; and, 'Native Title and Ecology: Political/legal transformations and sustaining ecologies', *Forty Years On: Political transformation and sustainability since the Referendum and into the future*, 5-8 November 2007, Canberra. I thank all the authors for their chapters, and their patience with the peer review and editing process. For their feedback on the draft collection, I thank the two anonymous peer reviewers. For their editorial assistance I thank Leah Ginnivan, Lydia Glick, Claire Stacey, Cynthia Ganesharajah and Geoff Hunt. For their help with The Australian National University committee process, I thank Rani Kerin, Peter Read and Lisa Strelein.

Country, Native Title and Ecology was undertaken as part of my Research Fellowship in the Native Title Research Unit, within the Indigenous Country and Governance Research Program at the Australian Institute of Aboriginal and Torres Strait Islander Studies (AIATSIS). AIATSIS acknowledges the funding support of the Native Title and Leadership Branch of the Department of Families, Housing, Community Services and Indigenous Affairs. The views expressed in this collection are the views of the authors.

List of Shortened Forms

AD   Avoided Deforestation

BAC   Bawinanga Aboriginal Corporation

CDEP   Community Development Employment Projects

CDM   Clean Development Mechanism

CFI   Carbon Farming Initiative

CSIRO   Commonwealth Scientific and Industrial Research Organisation

DCCEE   Department of Climate Change and Energy Efficiency

DSE   Department of Sustainability and Environment

IFIPCC   Intergovernmental Panel on Climate Change

ILC   Indigenous Land Corporation

ILUA   Indigenous Land Use Agreement

IPA   Indigenous Protected Area

IPCC   Intergovernmental Panel on Climate Change

MLDRIN The Murray Lower Darling Rivers Indigenous Nations

NCOS   National Carbon Offset Standard

NRM   Natural Resource Management

NTA   *Native Title Act 1993* (Cth)

RDA   *Racial Discrimination Act 1975* (Cth)

REDD   Reducing Emissions from Deforestation and Forest Degradation in Developing Countries

RNTBC   Registered Native Title Bodies Corporate

UNCBD   United Nations Convention on Biological Diversity

UNDRIP United Nations Declaration on the Rights of Indigenous Peoples

UNEP   United Nations Environment Program

UNFCCC United Nations Framework Convention on Climate Change

UNPFII   United Nations Permanent Forum on Indigenous Issues

WAI   Western Agricultural Industries

WALFAP West Arnhem Land Fire Abatement Project

WMO   World Meteorological Organisation

YCEC   Yirrkala Community Education Centre

Contributors

Hanson Boxer (Pampila) is a Walmajarri Elder who grew up near Fitzroy Crossing and now lives in Broome. His family had left Kaningarra in the Great Sandy Desert some 25 years before he was born. In 1989, together with his wife, Laurel, he set up Yakanarra Community on Old Cherrabun Station. He is also an accomplished sculptor, musician, and dancer.

Warford Bujiman (Pajiman) was brought in from the Great Sandy Desert by his parents when he was a young boy, and grew up on Cherrabun Station south of Fitzroy Crossing. He is now retired and living in Fitzroy Crossing.

Emily Gerrard is currently a Senior Associate at Allens Arthur Robinson, having previously worked as a lawyer at Native Title Services Victoria. Her experience includes advising on environment, planning and native title law, particularly in relation to energy and resources projects. Her practice also involves work in climate law in Australia and the Asia Pacific. Outside of her work at Allens, Emily continues to work with Indigenous communities across Australia through the National Indigenous Climate Change Project, an initiative devised and driven by Indigenous Australians to support their engagement in emerging economic development opportunities.

Lee Godden is a Professor at the Melbourne Law School and Director of the Centre for Resources, Energy and Environmental Law. Her research interests include: Indigenous peoples' land and resource rights, and the intersections with environmental law and natural resources management. She has conducted comparative research in these fields in Australia, Canada, New Zealand, South Africa, and Asia-Pacific, with a current research grant on climate change, environmental governance and customary land and resource rights.

Doug Moor (Kordidi) was a Walmajarri man who spent his life at Cherrabun Station and later was an Elder of the Djugerrari community. He is deceased.

Mervyn Mulardy Jnr is a Karrajarri man, whose country is south of Broome in the West Kimberley region of Western Australia. Mervyn is a young leader for his people. He is a cultural musician, artist, and the Chairperson of the Karajarri native title corporation. Mervyn has performed music and dance around the world, and was a major force within his own community's fight for native title. Mervyn is also a cultural adviser and founding member of the Yiriman Youth Project, and has represented Karajarri on the Executive Committees of the Kimberley Land Council and Kimberley Aboriginal Law and Culture Centre.

Samantha Muller is an Adjunct Fellow at Flinders University. Since 2001 she has been undertaking research with and working for independent Aboriginal and Torres Strait Islander organisations in north-east Arnhem Land, Torres Strait and Adelaide. She is the author of several journal articles and book chapters relating to Indigenous land and sea management in Australia, including discussion of the Indigenous Protected Areas program, accountability, Indigenous ranger programs and sea country.

Roy Stone is the Program Manager in charge of water accounting in the Department of Water, Western Australia. Roy graduated in Civil Engineering from the University of Western Australia and has more than 30 years experience in the water industry, primarily on water resource planning.

Patrick Sullivan is a Research Fellow at the Australian Institute of Aboriginal and Torres Strait Islander Studies and an anthropologist. He is also Adjunct Professor at the National Centre for Indigenous Studies (ANU). His main field of research is the engagement of Aboriginal people and organisations with the Australian public sector. He has worked in Kimberley region, West Australia, since 1983. Much of his professional life has been spent working with independent Aboriginal organisations. He is the author of numerous scholarly articles, as well as practical reports on heritage and land issues. His book *Belonging Together: Dealing with the Politics of Disenchantment in Australian Indigenous Policy* (2011) is published by Aboriginal Studies Press.

Luke Taylor is an anthropologist who specialises in research with Aboriginal and Torres Strait Islander artists and has written a number of books on Aboriginal art, including *Seeing the Inside: Bark Painting in Western Arnhem Land* (Clarendon Press, Oxford, 1996). He is the Deputy Principal, Research and Information at the Australian Institute of Aboriginal and Torres Strait Islander Studies and an Adjunct Professor with the Centre for Cross-Cultural Research at The Australian National University.

Jessica K Weir (Editor and Author) has published widely on water, native title and governance, and is the author of *Murray River Country: An Ecological Dialogue with Traditional Owners* (Aboriginal Studies Press, 2009). Jessica's work was recently included in Stephen Pincock's *Best Australian Science Writing 2011*. In 2011 Jessica established the AIATSIS Centre for Land and Water Research, in the Indigenous Country and Governance Research Program at the Australian Institute of Aboriginal and Torres Strait Islander Studies.

1. Country, Native Title and Ecology[1]

Jessica K Weir

The overtly technical process of making a native title application has obscured one of the central reasons why Indigenous people engage with the native title system – to affirm and promote their relationships with country. This publication focuses on Indigenous peoples' relationships with country, and seeks to discuss native title in terms that are more directly related to those relationships. In doing so, we also describe ways of living on country that inform and critique mainstream land and water management. This volume also includes case studies that are not classified as part of the native title system, so as to broaden native title issues into the frame of traditional ownership. Limitations with common and statutory native title law have meant that native title is not a land justice system accessible to all traditional owners of country. Profound connection to country frequently exists where native title cannot be successfully applied for, or where traditional owners choose not to make native title applications.

The focus in this volume on native title *and* ecology is made because many traditional owners express their attachment for country through their unique ecologies, and the discipline of ecology's focus on relationships links into the holistic language of country. Framing native title with ecological thinking also deliberately challenges the contributing authors to resist being led by the law, culture and bureaucracy of native title. Instead, we bring our attention to issues of land and water, the relationships Indigenous people hold with species and their ecologies, and the challenges and opportunities of native title to sustain life.

Currently, academics are rethinking their disciplines and knowledge traditions in response to environmental devastation, to consider why we are not better able to value and respond to our ecological relationships. The wider significance of this work is to consider how we should live in this time of global climate change and widespread environmental destruction. In Australia, this intellectual rethink is informed by Indigenous peoples' knowledge about country. Indigenous people have knowledge systems and concepts, territories, cultures and customary activities centred on living with country. In recent times, this inheritance has been interacting with the laws, knowledges, cultures and customs introduced by non-Indigenous people. This intercultural context highlights the different

1 I would like to thank Luke Taylor, Tran Tran and two anonymous peer reviewers for their comments on an earlier version of this paper. For their editorial assistance I thank Leah Ginnivan, Lydia Glick, Geoff Hunt, Claire Stacey, and Cynthia Ganesharajah.

and similar cultural concerns that Indigenous and non-Indigenous people bring to intellectual frameworks that prioritise the life of country, life which inspires all people to live on and enjoy country.

Countries and ecologies

Country is a term many Aboriginal people use generally to describe their traditional lands and waters, although it has a much broader meaning than just territory. Riley Young spoke to Deborah Rose about the importance of his country, which includes the Yarralin community in the north of the Northern Territory:

> We been borning [in] this country. We been grow up [in] this country. We been walkabout this country. We know all this country all over … Blackfellow been born top of that ground, and blackfellow-blackfellow blood [in the ground] … This ground is mother. This ground, she's my mother. She's mother for everybody. We born top of this ground. This [is] our mother. That's why we worry about this ground. (R Young, cited in Rose, 2000 [1992]: 220)

Country can be one's own campfire, one's family area, a clan area, a geographical region, a language area and, or, an ecological zone. Country might be similar to a drainage basin for a river, or be marked by a particular plant species, or not. It depends on the context (Rose, 2000 [1992]: 117). Country is much more dynamic than a people-place match; it is multilayered and indeterminate with multiple forms and meanings (Sutton, 1995: 49–50). One person will have multiple countries, and one country will have multiple people. Because people have multiple affiliations to different countries, they will always be negotiating these relationships over their lifetime.

Many Indigenous people speak about the importance of their country because of who they are and how their way of life is embedded in their country. They continue to live on their country or maintain their links to country through visits and other means. There are also Indigenous people seeking to reconnect with country, including some members of the Stolen Generations. Many Indigenous people live in communities, towns and cities on the country of other people. They are sometimes known as 'historical peoples', in that they moved, or were moved, historically to those lands and have chosen to stay and invest in the new community.

Country is useful to examine as a philosophy of existence. In its most expansive sense it is where knowledge comes from. Country is where the rules for existence and many of the relationships between species and humans were established by ancestral creative beings (Rose, 2000 [1992]: 43–44). In the Dreaming creative beings travelled over the land creating species and geographical features. The Dreaming is not past or present but 'everywhen' (Stanner, 1989 [1963]: 228). Rose describes what she calls the Dreaming ecology:

> Everything comes out of the earth by Dreaming; everything knows itself, its place, its relationships to other portions of the cosmos. Every living thing has, and knows, its own Law. (Rose, 2000 [1992]: 220)

Knowledge, country, species and people are co-created. Country is the locus of this knowledge and ecological life.

Critically, in country humans and nature, and nature and culture, are not regarded as separate, but are entangled together in all types of relationships. Species and their relationships with humans are spoken about in terms of language, law, religion and stories (see, for example, Kinnane, 2002; Langton, 2002; Rose, 1996; Smith, 2005). This knowledge teaches which relationships are important and why, with a focus on life sustaining networks that can also be called 'connectivity' (Weir, 2008). This is an expanded connectivity that goes beyond food and energy webs to encompass stories, histories, feelings, shared responsibilities and respect.

Ecologists use the term connectivity in a narrower sense to describe the way animals and plants are interconnected through energy webs with each other, across places and through time. Ecologists recognise that each species has its own understanding of how to survive where they live, and ecologists call this their ecological niche or habitat (Allen and Hoekstra, 1992: 169; Manning *et al.*, 2004: 622–623). Importantly, ecology has been part of the reinvigoration of science towards more integrated thinking about environmental issues (Worster, 2008 [1997]; see also, Godden, this volume). Ecologists are 'subversive scientists' in their rejection of the atomism of reductive laboratory work (Worster, 2008 [1997]: 22). Ecologists engage with organic complexity and connection within and across ecosystems.

Ecology is often described as a sub-discipline of biology, and the subjectivity of this approach is evident in its history and contemporary context. Biology and ecology are influenced by the 'balance of nature' assumption arising out of seventeenth century Christian theology. In the twentieth century, ecologists debunked the idea of nature as equilibrium, with plants and animals in stable populations, and replaced it with a dynamic focus on energy flow, and plant and animal productivity and succession (Egerton, 1973: 330). Donald Worster has followed the evolution of theories about ecosystems, from natural order to complexity, chaos and disorder (Worster, 2008 [1997]: 412). Worster has also described how ecologists find themselves propelled into environmental activism because of the huge rise in species extinctions (2008 [1997]: 419). Together with conservation biologists, ecologists have been at the forefront of documenting and responding to the rapid loss of biodiversity that is occurring in contemporary times (for example Lindenmayer, 2007).

One of the most powerful legacies of the western scientific tradition has been the separation of humans from nature, which developed as a central tenet of the natural sciences. In the conservation tradition this separation is also known as 'wilderness thinking', where a pristine nature exists outside of human activity (Braun, 2002: 2). Wilderness thinking has been replicated in western land tenure systems that separate nature into definable people-free spaces, including national parks, reserved lands, and protected areas. In Australia, these land tenures have excluded Indigenous people from their country. The erasure of Indigenous peoples' rights complicit in these land tenures continues to have ramifications in native title, limiting the kind of native title rights that can be recognised (see discussion below). Human rights infringements resulting from wilderness thinking have forced a rethink of this ideology, to create participatory conservation practices (Campese *et al.*, 2009). In Australia, Indigenous people are now accessing conservation monies as part of managing their native title lands and waters.

The influence of separating humans from nature has also contributed to a rationalist and utilitarian approach to country (Kinnane, 2002: 24). Nature has become natural resources, simplified as matter that is external to humans and for human consumption (Braun, 2002: 41; Robin, 2007: 186). Combined with the achievements of the industrial revolution, separation thinking has enabled proponents to transform nature on a massive scale, without regard for the delicate web of relations which bind together all creatures (Latour, 2001 [1993]: 32, 39). In the academic response to ecological devastation, ecologists, anthropologists, geographers, historians, eco-philosophers, lawyers and others are re-situating humanity within nature, and extending subjectivity to nature (see, for example, Graham, 2011; Ingold, 2000; Latour, 2001 [1993]; Manning *et al.*, 2004; Mathews, 1994; Mitchell, 2002; Robin, 2007; Rose, 1996). From this position, we are not taking a view of the world, but 'taking a view *in* it' (Ingold, 2000: 42). Within this world, we acknowledge our lives are in connection with multitudes of other beings, and we can better foreground these ecological relationships in our decision making. This disciplinary re-think is both a response to separation thinking, and a move towards more holistic or integrated thinking. This is also a move towards the knowledge tradition of country and an expanded connectivity.

Country and ecology relate as knowledge traditions because they explicitly acknowledge the interdependence of life, albeit these are expressed differently. Because Indigenous people have an expanded notion of connectivity, they make connections that cannot be categorised as just an energy exchange. For example, Indigenous people repeatedly identify the role of water as a life force, but link that to ceremony, song, protocols and survival (Yu, 1999). In Arnhem Land, major regional ceremonies ensure the arrival of the wet season, and thus the good health of the participants through the multiplication of species and continued opportunities for good hunting (Taylor, this volume). Walmajari people in south west Kimberley also conduct ceremonies to ensure the rains come, linking the rains with invisible seeds that produce the animal species associated with water, such as goanna, frogs, land crabs, fresh water eels, turtle, fish, and ducks (Sullivan and others, this volume). Conversations about the unique ecologies and species of particular places provide an important site of engagement between Indigenous people and ecologists. In specific places there are many unique plants and animals, which are described as endemic by ecologists. Indigenous peoples' often very detailed and intimate knowledge about these species is based in thousands of years of observation, experimentation and teaching (Horstman and Wightman, 2001: 99).

Indigenous peoples' ecological knowledge is sometimes described in western scientific discourses as 'ethnoscience' or 'non-science', and treated as inventory knowledge similar to eighteenth century botany (Braun, 2002: 307, fn 1). This hierarchical relationship between knowledges is aided by the presumed universal approach of western science. By presuming universalism, western science challenges the validity of other local knowledge systems, whilst denying its own cultural origins in Euro-American capitalist understandings of the world (Sillitoe, 2007: 13). As described above in relation to ecology, all 'sciences' have socio-political, cultural and historical contexts that inform how knowledge is prioritised, sought and interpreted. Understanding the cultural context of different sciences is important so that Indigenous peoples' knowledges are not described (and then dismissed) as 'cultural'; that is, as context specific and not capable of communicating in and across other contexts.

The positioning of Indigenous peoples and their knowledges as 'cultural' or 'traditional' within (presumed universal) Eurocentric knowledge frameworks has implications not just for understanding their ecological knowledge, but also for the recognition of their rights. Narrow legal interpretations of 'tradition' have had a profound effect on who is recognised as having native title, and what those native title rights are. Because of the close relationships between country and ecology, ecological considerations are either explicitly or implicitly part of these encounters.

Native titles

Native title recognises Indigenous peoples' laws, customs and connections with their lands and waters. Native title was recognised in the 1992 High Court *Mabo* decision (*Mabo v Queensland [No 2]*, 1992), although the Court had previously acknowledged the distinct legal rights of Indigenous peoples in country.[2] Following *Mabo*, native title was enacted in Federal legislation as the *Native Title Act 1993* (Cth) (NTA). The NTA establishes a system for recognising native title, making native title applications, forming native title agreements, and for holding and managing native title rights. Native title rights and interests are unique for each native title holding group, as based in their laws and customs, and reflecting the diversity of Indigenous peoples' cultural, legal and political traditions. However, native title is not the same as those laws and customs, rather it is the recognition of them. In this recognition, native title brings with it a sweep of intercultural interactions, intricacies and ideologies around the law, Indigenous peoples' cultures and traditions, and separation thinking and connectivity thinking.

The ideology of 'tradition' has played a large role in the recognition of native title, Indigenous peoples' rights, and their intersection with issues of ecology and economy. Native title statutory and common law have defined native title as traditional. Indigenous peoples changes to and adaptations of their laws and customs are permissible so long as the laws and customs remain traditional enough. For example, whether Indigenous people use a gun or a spear to go hunting is not at issue, but whether they identify as a society of people with traditional laws and customs is. This emphasis on tradition is fraught because traditions are socially constituted in both contemporary place and time. Whilst the defining feature of traditions is a reference to continuity with the past, traditions are constantly constructed and reconstructed to serve contemporary purposes (Otto and Pedersen, 2005: 11, 31).

The most well known native title determination that hinged on different interpretations of tradition is *Yorta Yorta* (*Yorta Yorta Aboriginal Community v the State of Victoria and Ors*, 1998, unreported). This decision related to Yorta Yorta contemporary expressions of looking after country, and their participation in economic life. Yorta Yorta country is river country on the border of southern New South Wales and northern Victoria. At their native title trial, Yorta Yorta talked about cultural heritage and national resource management as part of being traditional owners of country (Muir and Morgan, 2002: 5). Another part of their evidence of connection was an 1881 petition requesting the Governor of New South Wales for some of their traditional country so that they could become

---

2   With respect to Gunditjmara people, in *Onus v Alcoa of Australia Ltd* (1981) 149 CLR 27 (Weir, 2009b: 13).

independent farmers. However, in the Federal Court decision, Justice Olney determined that the Yorta Yorta had 'abandoned' their native title traditions, and that environmental conservation was not Yorta Yorta culture (paras 126, 128). The adoption of commercial farming by the Yorta Yorta was antithetical to their status as traditional owners (Strelein, 2009: 75). Olney found that the Yorta Yorta's traditional laws and customs had been 'washed away by the tide of history'.[3] Such is the influence of 'tradition' on Indigenous identity, that Indigenous people who seek to make a commercial livelihood from country, as the Yorta Yorta petitioned for in 1881, are not only being untraditional, they are not eligible for Indigenous rights, and their status as an authentic Indigenous person is challenged (Weir, 2009a: 23).

It is native title's narrow understanding of tradition which frames native title as uneconomic (Strelein and Weir, 2009). Whilst there are a few exceptions, in general the lists of native title rights and interests recognised in native title determinations do not include economic or commercial rights and interests. Aboriginal people somehow exist outside the modern economic space. Their lived reality is very different, including their native title work. For example, native title holders negotiate access to their lands and waters with mining and exploration companies in a very commercial context. Indigenous people leverage their right to negotiate for economic benefits for their native title corporation (O'Faircheallaigh, 2007).

The debate on being traditional and being Indigenous engages with ecological issues through western notions of Indigenous people as noble savages living in harmony with nature, and untouched by civilisation (Hames, 2007). Indigenous people are exempted from the nature-society hyper-separation, and instead collapsed into nature as part of the fauna and flora (Braun, 2002; Langton, 1995). The simplistic matching of Indigenous and conservation agendas is evident in environmental forums when Indigenous people either exploit this notion, or challenge it as being prejudicial. These tensions have been evident in the debate over the *Wild Rivers Act 2005* (Qld), which has placed environmental protection measures on rivers in Cape York, and the impact of this legislation on native title rights and interests. This is also evident in the decision by Woodside Energy Ltd to locate a gas processing facility at Walmadan (James Price Point) north of Broome in the Kimberley, which will bring industrial development to areas of immense natural and cultural heritage. The Kimberley and Cape York have a history of collaborations between environmental interests and Indigenous interests, but these working relationships have unraveled as expectations and priorities differ around Indigenous rights and conservation. Industrial capitalism has perpetuated the misalignment of the motivations of Indigenous people

---

3   *Yorta Yorta Aboriginal Community v the State of Victoria and Ors*, 1998, unreported, para 19. This phrase was used repeatedly in the determination, see paras 3, 126, 129.

and environmental groups, by entrenching ecological and economic priorities as oppositional (Weir, 2009a: 24-25). This relationship is being rethought in sustainability studies to identify synergies between economic, ecological and social goals.

In addition to ideologies of tradition, native title is strongly influenced by interpretations of statutory and common law. Native title is an intensely legalistic system, which has its own technical rules and self-reinforcing legal narrative (Ritter, 2010: 193). This governance context can have unpredictable outcomes. For example, the recognition of native title on national park and other reserved land tenures differs substantially between State and Territory jurisdictions (see also Bauman and Haynes, forthcoming). In New South Wales and Queensland non-exclusive native title rights have been recognised. However, in the 2002 High Court *Ward* (*Western Australia v Ward,* 2002) decision for the Miriuwung and Gajerrong peoples, the Court found that native title rights did not survive the vesting provisions of most national parks in Western Australia (Strelein, 2009: 70). In the same *Ward* decision, the High Court found the opposite with respect to national parks in the Northern Territory. In the Territory, problems with the tenure and statute for Keep River National Park meant that its title was void – bringing the validity of all national parks in the Northern Territory into question. The Northern Territory government responded with a comprehensive agreement-making process to set up joint-management provisions with traditional owners for the majority of the Territory's national parks, irrespective of whether native title had been recognised at each particular park (Dillon and Westbury, 2007: 96–111). In Western Australia, traditional owners and native title holders have had to lobby for joint-management legislation to overcome the discriminatory effect of the state legislation. This is now starting to be addressed through amendments to State conservation legislation.[4]

The *Yanner* (*Yanner v Eaton,* 1999) case provides another example of how native title court decisions are renegotiating environmental management in relation to the distinct legal rights of Indigenous peoples. In late 1994 Murrandoo Yanner killed two young estuarine crocodiles from Cliffdale Creek, which is in his country near the Gulf of Carpentaria in Queensland. He ate the crocodile meat together with other members of his clan. Estuarine crocodiles are a protected species under the *Fauna Conservation Act 1974* (Qld). They are protected in Queensland because the crocodiles had been hunted to near extinction for their skins. Yanner was charged for breaching the *Fauna Conservation Act,* his defence was that he had native title rights to kill and eat the crocodiles. In 1999, the High Court upheld Yanner's native title rights. The justices determined

---

4   The Conservation Legislation Amendment Bill 2010 will amend the *Conservation and Land Management Act 1984* to enable joint management arrangements between the Department of Environment and Conservation and other landowners, including Aboriginal people.

that the government cannot own wildlife, it can only regulate wildlife and this regulation is not enough to extinguish native title. Yanner's native title rights to hunt crocodiles for domestic purposes had survived this regulation.

Both *Ward* and *Yanner* highlight how Indigenous peoples' native title rights and environmental policy intersect. Native title rights often survive on reserved lands, and Indigenous people have distinct rights to hunt, fish and gather native species. Environmentalists have lobbied for species protection and the reservation of protected areas as part of governments' responsibilities to respond to environmental issues such as the clearing of native forests and the loss of species diversity. Indigenous peoples' rights with respect to such lands and species are distinguished as distinct because they are distinctly different from other people in Australia – they have their own laws, traditions and customs in relation to Australia's lands and waters (Weir, 2009a: 115–116).

Beyond the legal framing of native title, *Mabo* has delivered a shift in societal attitudes to better recognise the distinct relationships between traditional owners and their country. For Yorta Yorta, despite their native title experience they continue to be recognised as the traditional owners of country. The Victorian government opposed their native title application, but soon after they signed a joint-management agreement with Yorta Yorta over the river red gum forest in the centre of Yorta Yorta country (Atkinson, 2004).

Native title as environmental management

The recognition of traditional owners in Australian society and government is delivering a greater involvement of Indigenous people in ecological issues, often through environmental partnerships with governments on native title, land rights, and public lands more generally.

Native title is increasingly part of the environmental management picture for Australia because of the sheer extent of its recognition. Over 1,195,935 square kilometres of land has been determined as native title, or about 15 per cent of Australia's land mass, and representing around 100 determinations.[5] In addition, there remains a backlog of more than 350 native title applications awaiting determination by the courts. Much native title recognition has been in northern and 'remote' Australia, where native title determinations have been as large as 136,000 square kilometres in the case of the Martu people. In remote Australia native title is often recognised as exclusive possession. Determinations in settled

5   From 'Determinations of Native Title: as at 30 June 2011', <http://www.nntt.gov.au/Publications-And-Research/Maps-and-Spatial-Reports/Documents/Quarterly%20Maps/Determinations_map.pdf> (accessed 30 August 2011).

Australia are often non-exclusive possession, and recognised on much smaller pieces of land because of the land tenure history. Together with land rights lands and other Indigenous land holdings, this 'Indigenous Estate' comprises almost 20 per cent of Australia (Altman, Buchanan, Larsen, 2007).

The extent of the Indigenous Estate means that strategies designed to respond to national environmental issues must engage with native title holders and traditional owners. This includes climate change adaptation and mitigation, our priorities for water, and our management of invasive pests, diseases and weeds. Native title land holdings also have implications for the environmental management of particular places, which is nationally important as they include many areas of high conservation and biodiversity significance (Altman *et al.*, 2007: 14). Collaboration is central where non-exclusive native title possession is recognised on public lands, such as national parks and reserved lands, state forests, stock routes, and pastoral lease lands. Where exclusive native title is recognised, native title holders arguably carry the responsibilities of land holders as deemed under the various land acts of the States and Territories.[6]

There are several native title governance bodies that Indigenous people work with as part of applying for and holding native title. Native title applicants work closely with Native Title Representative Bodies and Native Title Service Providers to bring together evidence of their native title, as based in their traditional laws and customs, and to reach a native title determination either through litigation or mediation. If native title is recognised, the native title holders are required by the NTA to set up a Registered Native Title Body Corporate (RNTBC) to manage their native title rights and interests, and to provide a legal entity for other people who wish to do business on native title land (AGDSC 2006; Weir, 2007). There are now about 80 RNTBCs in Australia.[7] These RNTBCs are the key native title governing body for native title lands and waters, including land and water issues, community services, economic development, and more. They continue to receive assistance from Native Title Representative Bodies and Native Title Service Providers in matters such as native title agreements (known as Indigenous Land Use Agreements) and future act negotiations. Future acts are procedural rights to be notified, consulted or involved in negotiations with respect to certain development activities. A future act could be the development of a mine, a house or a national park.

---

6   This legal position is currently being examined in a research project into weeds management on native title lands, which is being undertaken at the Australian Institute of Aboriginal and Torres Strait Islander Studies, funded by the Rural Industries and Research and Development Corporation.

7   RNTBCs are also known as PBCs – Prescribed Bodies Corporate, as this is the name they are known by prior to their registration after a successful native title determination.

RNTBCs do not receive operational funding, and are advised to seek funding from government and private grant monies (AGDSC, 2006). This has presented numerous challenges for RNTBCs (Bauman and Tran 2007; Bauman and Ganesharajah 2009; Weir, 2011), however an important although limited source of funds is grant monies for environmental management. For example, the Federal Government's Working on Country program provides funding for Indigenous people to work as rangers on their own land, water and sea management projects. There is also an emerging trend for collaborations with the Federal Government's Indigenous Protected Area program. The Indigenous Protected Area program establishes reserved lands governed by Indigenous people on Indigenous held and some public lands (Bauman and Smyth, 2007). In exchange for resources such as staff wages and a vehicle, native title holders agree to certain environmental protections and outcomes on their land. The resources resulting from these collaborations are of immense value for under-resourced RNTBCs, who are sometimes also supported by land and sea units based within Native Title Representative Bodies. However, matching native title with environmental land tenures may have unintended consequences, including perhaps limiting their eligibility to participate in carbon economies (Gerrard, this volume).

Both the Indigenous Protected Area and Working on Country programs recognise the importance of Indigenous peoples' land holdings for environmental management, as well as Indigenous peoples' 'caring for country'. Caring for country is a term used to describe Indigenous peoples' land and sea management. As discussed above, there is a reciprocal relationship between people and country. This is reflected in the familiar saying by Indigenous people that 'if you look after the country, the country will look after you' (Griffiths and Kinnane, 2010: iii, 3). This is supported by research into the health and other benefits of caring for country activities (Burgess *et al.*, 2005; Garnett and Sithole, 2007). Government literature from the Indigenous Protected Area and Working on Country programs also relate caring for country to broader outcomes in terms of supporting Indigenous people to live on country, with intergenerational benefits such as keeping culture strong, meaningful economic opportunities, health, education and social cohesion (DEWHA, 2009; DEWR, 2007).

The importance of environmental jobs in enabling traditional owners to continue to live on country is discussed by Luke Taylor (Chapter 2, this volume). Taylor draws on conversations with Kuninjku language speaker Ivan Namirrkki, to describe one man's connection to his country, an outstation in a stringybark forest in western Arnhem Land. Taylor relates Namirrkki's appreciation of country to Kuninjku concepts of personhood, sociality, power, health, and aesthetic experience. Namirrkki's outstation is in the country of another Kuninjku clan, not his own, and his negotiated rights to live here are an acknowledgement of his personal historical circumstances that led to his knowledge of this country. It is a good example of the dynamic contemporary relationships between people and country. Yet Namirrkki's outstation life is recast as anachronistic in national debates framed by the participation of Aboriginal people in the national economy. Taylor counters this with examples of Aboriginal people taking advantage of the on-country enterprises available to them. Through these kinds of work effort and economic opportunities, Aboriginal people continue to articulate their concerns to nurture country whilst also connecting with western modes of action and thinking.

Patrick Sullivan, and co-authors Hanson Boxer (Pampila), Warford Bujiman (Pajiman), and Doug Moor (Kordidi) go so far as to say that environmental collaborations are perhaps more important than native title recognition (Chapter 3, this volume). The authors describe the importance of water to Walmajari people of the Great Sandy Desert. These waters are living waters, intimately linked with the *kalpurtu* who are the very first beings. The importance of the wet season in providing food and water is evident in the energy Walmajari invest in rain making ceremonies. The authors describe the comfort, homeliness and sociality of various billabongs and soaks, used cyclically by people through the seasons. This is a holistic cultural system which combines spiritual practices and the use of local knowledge to look after country. The authors question whether native title could do justice to these beliefs, or protect their connection to their local water sources which have suffered environmental degradation from pastoral activities. They note that native title recognition itself does not provide the necessary resources and organisation to revitalise the land. Instead, what is needed is a resourced 'two-way' alignment of priorities between settler and Aboriginal stakeholders, to introduce conservation measures and renew cultural activities.

Yolngu people from north-east Arnhem Land have developed the term 'two ways management' as a counterpoint to the dominance of non-Indigenous science in natural resource management projects. Samantha Muller discusses the Yolngu approach with respect to the Yellow Crazy Ant Eradication Program, a project which Yolngu people have initiated in partnership with various government and non-government participants (Chapter 4, this volume). Two-ways management is intended to be collaborative across Indigenous and non-Indigenous ways of knowing country. But in the Crazy Ant project, ontological, logistical, cultural and fiscal challenges thwart this intention. At base, there are different perspectives on the project itself; whether it is just about an ant, or about the meeting and exchange of two knowledge systems. The non-Indigenous senior scientist admits that he cannot see the relationship between an invasive introduced pest and Yolngu knowledge. Muller argues that to improve the legibility of ideas across cultures we need to better develop a *lingua franca* – a common language – to ensure that such partnerships are meaningful for both parties. Muller also identifies the lack of equal resourcing as a key inhibiting factor in these culturally rich environmental collaborations.

Finding common ground in environmental issues is a theme in Chapter 5, written by myself, Roy Stone and Mervyn Mulardy Jnr. The Chapter documents an engagement between native title, water management, and a large scale cotton proposal in the West Kimberley south of Broome. This development proposal prompted research into the different values held in groundwater that had been identified for future agricultural consumption. What is unique about this water planning research is the methodology. The joint cultural and ecological fieldwork revealed the common ground held between Indigenous and hydro-ecological knowledge, thereby facilitating the immediate relevance of Karajarri water knowledge to contemporary water management and planning. This relevance is not reflected in Karajarri native title. Karajarri provided their intimate knowledge of groundwater as native title evidence, but their consent determination did not include native title rights or interests to that water.

Lee Godden (Chapter 6, this volume) considers the growing trend of agreement making with Indigenous people, and how such agreements co-locate native title and ecology within legal, economic and social framings. Agreement making has been adopted across the native title system as a way of managing the problems with the rules of native title, as formalised in Indigenous Land Use Agreements. Godden relates this agreement making trend to wider structural changes precipitated by global processes and ideological influences. One of these structural changes is the movement of governments away from delivering goods and services, towards using contracts to govern many areas for political, social and economic life. Often ecology is explicitly manifested in these agreements as joint-management, but it is more common that ecological concerns are diffused through the agreement, such as the stewardship responsibilities that come with native title recognition over pastoral leases. Godden considers how these environmental collaborations are currently being reworked through the increasing use of market tools to address environmental issues, and is concerned that ecology and equity are being displaced by the economic context of efficiency and development.

This 'market environmentalism' is part of the reinvention of conservation using market tools, and perhaps its best known example is carbon trading in response to climate change. Carbon trading is a key strategy to reduce the increase of greenhouse gases in the atmosphere, and thereby mitigate the effects of climate change. In Australia, climate change will alter local ecologies as rain patterns change, sea levels rise, and plant and animal species move south. In the north, the potential for increased rain and cyclonic activity will challenge living conditions in remote communities (Green, Jackson and Morrison, 2010). Climate change is not just changing how we frame our environmental priorities, but has brought ecological considerations into the centre of industrial and political agendas.

Emily Gerrard (Chapter 7, this volume) identifies that Indigenous people have a special interest in climate change because of their unique relationships with country, and their specialised knowledge. However, climate change related laws, regulations and markets have the potential to further decrease or limit Indigenous peoples' rights and interests in country and its resources, either through the extinguishment of native title or restricting access rights to land and resources. Gerrard argues that the international principle of 'common but differentiated responsibility' has relevance here in a national context. This principle encourages a shared responsibility to climate change while protecting certain populations from a disproportionate burden of meeting mitigation and adaptation obligations. Gerrard makes it clear that this substantive equality approach to climate change requires action in terms of legal foundations and policy incentives to support Indigenous people to be responsive to the carbon constrained world.

There are economic opportunities in market environmentalism, conservation, environmental management, and so forth for on-country enterprises, which will support the viability of Indigenous people living and working on country. But climate change and ecological degradation are not just economic opportunities, they are a wake-up call to rethink how we can live on this planet. Human activities have become so influential on ecosystems that they define our current geological epoch.

Part of ensuring that Australia's land and water management regimes recalibrate to sustain ecological relationships is to learn from and work with Indigenous people. However, it has taken a long time for law and policy to begin recognising Indigenous peoples' governance arrangements, legal rights, and intellectual approaches to living with and looking after country. The chapters in this book give examples of the knowledge and approaches traditional owners bring to country, and how to better manage the interface between Indigenous and non-Indigenous epistemologies and ways of relating to country. Contentions over meaning, knowledge, and authority will persist, but they should not prevail to undermine the goal of sustaining life.

References

Allen, TFH and TW Hoekstra 1992, *Toward a Unified Ecology*, Columbia University Press, New York.

Altman, J, G Buchanan and L Larson 2007, *The Environmental Significance of the Indigenous Estate: Natural Resource Management as Economic Development in Remote Australia*, CAEPR Discussion Paper No. 286/2007, Centre for Aboriginal Economic Policy Research, Australian National University, Canberra.

Atkinson, H 2004, 'Yorta Yorta Co-operative Land Management Agreement: impact on the Yorta Yorta Nation', *Indigenous Law Bulletin* 6(5): 23–25.

Attorney General's Department Steering Committee (AGDSC) 2006, *Structures and Processes of Prescribed Bodies Corporate*, Attorney General's Department, Canberra.

AGDSC – *see* Attorney General's Department Steering Committee

Bauman T and C Ganesharajah 2009, Second National Prescribed Bodies Corporate Meeting: Issues and Outcomes, Melbourne 2 June 2009, Native Title Research Report, Australian Institute of Aboriginal and Torres Strait Islander Studies (AIATSIS), Canberra.

Bauman, T and C Haynes (in review), 'Joint management and native title in Australian conservation areas: process, structure and partnership', AIATSIS Discussion Paper.

Bauman, T and D Smyth 2007, *Indigenous Partnerships in Protected Area Management in Australia: Three Case Studies*, AIATSIS in association with the Australian Collaboration and the Poola Foundation (Tom Kantor fund).

Bauman T and T Tran 2007, First National Prescribed Bodies Corporate Meeting: Issues and Outcomes, Canberra 11-13 April 2007, Native Title Research Report, AIATSIS, Canberra.

Braun, B 2002, *The Intemperate Rainforest: Nature, Culture, and Power on Canada's West Coast*, University of Minnesota Press, Minneapolis and London.

Burgess, P, FH Johnston, DMJS Bowman and PJ Whitehead 2005, 'Healthy country: healthy people? Exploring the health benefits of Indigenous natural resource management', *Australian and New Zealand Journal of Public Health* 29(2): 117.

Campese, J, T Sunderland, T Greiber and G Oviedo (eds) 2009, *Exploring Issues and Opportunities in Rights Based Approaches to Conservation*, CIFOR, IUCN and CEESP, Bogor, Indonesia: 123–140.

Department of the Environment, Water, Heritage and the Arts (DEWHA) 2009, *Working on Country – A Retrospective 2007 – 2008,* Department of the Environment, Water, Heritage and the Arts, Canberra.

DEWHA – *see* Department of the Environment, Water, Heritage and the Arts

Department of Environment and Water Resources (DEWR) 2007, *Growing Up Strong: The First 10 years of Indigenous Protected Areas in Australia,* Department of the Environment and Water Resources, Canberra.

DEWR – *see* Department of Environment and Water Resources

Dillon, C and ND Westbury 2007, *Beyond Humbug: Transforming Government Engagement with Indigenous Australia*, Seaview Press, West Lakes, South Australia.

Egerton, FN 1973, 'Changing concepts of the balance of nature', *Quarterly Review of Biology* 48: 322–350.

Garnett, S and B Sithole 2007, *Sustainable Northern Landscapes and the Nexus with Indigenous Health: Healthy Country, Healthy People,* Land and Water Australia.

Graham, N 2011, *Lawscape: Property, Environment, Law*, Routledge, Abingdon, United Kingdom.

Green, D, S Jackson and J Morrison (eds) 2010, *Risks from Climate Change to Indigenous Communities in the Tropical North of Australia*, Department of Climate Change and Energy Efficiency, Canberra.

Griffiths, S and S Kinnane, 2010, *Kimberley Aboriginal Caring for Country Plan – healthy country, healthy people*, report prepared for the Kimberley Language Resource Centre, Halls Creek.

Hames, R 2007, 'The ecologically noble savage debate', *Annual Review of Anthropology* 36: 177–190.

Horstman, M and G Wightman 2001, 'Karpati ecology: recognition of Aboriginal ecological knowledge and its application to management in north-western Australia', *Ecological Management and Restoration* 2(2): 99–109.

Ingold, T 2000, *The Perception of the Environment: Essays in Livelihood, Dwelling and Skill*, Routledge, London.

Kinnane, S 2002, 'Recurring visions of Australindia', in *Country: Visions of Land and People in Western Australia*, A Gaynor, M Trinca and A Haebich (eds), Western Australian Museum, Perth.

Langton, M 1995, 'Arts, wilderness and terra nullius', in *Ecopolitics IX: Perspectives on Indigenous People's Management of Environment Resources*, R Sultan (ed), Northern Land Council, Darwin.

— 2002, 'The edge of the sacred, the edge of death: sensual inscriptions', in *Inscribed Landscapes: Marking and Making Place*, B David and W Meredith (eds), University of Hawaii Press, Hawaii.

Latour, B 2001 [1993], *We Have Never Been Modern*, Harvard University Press, Cambridge.

Lindenmayer, DB 2007, *On Borrowed Time: Australia's Environmental Crisis and What We Must Do About It*, Penguin Books (in association with CSIRO Publishing), Camberwell.

O'Faircheallaigh, C 2007, 'Unreasonable and extraordinary constraints: native title, markets and the real economy', *Australian Indigenous Law Review* 11(3): 18–42.

Manning, AD, DB Lindenmayer and HA Nix 2004, 'Continua and Umwelt: novel perspectives on viewing landscapes', *Oikos* 104(3): 621–628.

Mathews, F 1994, *The Ecological Self*, Routledge, London.

Mitchell, T 2002 *Rule of Experts: Egypt, Techno-politics, Modernity*, University of California Press, Berkeley and Los Angeles.

Muir, J and M Morgan 2002, 'Yorta Yorta: the community's perspective on the treatment of oral history', in *Through a Smoky Mirror: History and Native Title*, M Paul and G Gray (eds), Aboriginal Studies Press, Canberra.

Otto, T and P Pedersen 2005, 'Disentangling traditions: culture, agency and power', in *Tradition and Agency: Tracing Cultural Continuity and Invention*, T Otto and P Pedersen (eds), Aarhus University Press, Denmark.

Plumwood, V 2002 [1993], *Feminism and the Mastery of Nature*, Routledge, London and New York.

Ritter, D 2010, 'The ideological foundations of arguments about native title', *Australian Journal of Political Science* 45(2): 191–207.

Robin, L 2007, *How a Continent Created a Nation*, University of New South Wales Press, Sydney.

Rose, DB 1996, *Nourishing Terrains: Australian Aboriginal Views of Landscape and wilderness*, Australian Heritage Commission, Canberra.

— 2000 [1992], *Dingo Makes Us Human; Life and Land in an Australian Aboriginal Culture*, Cambridge University Press, Cambridge.

Sillitoe, P 2007, 'Local science vs. global science: an overview', in *Local Science vs. Global Science: Approaches to Indigenous Knowledge in International Development*, P Sillitoe (ed), Berghahn Books, New York.

Smith, BR 2005, '"We got our own management": local knowledge, government and development in Cape York Peninsula', *Australian Aboriginal Studies* 2005(2): 4–15.

Stanner, WEH (ed) 1989 [1963], *On Aboriginal Religion*, University of Sydney, Sydney.

Strelein, L 2009, *Compromised Jurisprudence: Native Title Cases since Mabo*, 2nd edition, Aboriginal Studies Press, Canberra.

— and JK Weir 2009, 'Conservation and human rights in the context of native title in Australia', in *Exploring Issues and Opportunities in Rights Based Approaches to Conservation*, J Campese, T Sunderland, T Greiber, and G Oviedo (eds), CIFOR, IUCN and CEESP, Bogor, Indonesia.

Sutton, P 1995, *Country: Aboriginal Boundaries and Land Ownership in Australia*, Aboriginal History Monograph no 3, The Australian National University, Canberra.

Weir, JK 2007, 'Native title and governance: The emerging corporate sector prescribed for native title holders', *Land, Rights, Laws: Issues of Native Title* 3(9): 1–16.

— 2008, 'Connectivity', *Australian Humanities Review* 45: 153–164.

— 2009a, *Murray River Country: An Ecological Dialogue with Traditional Owners*, Aboriginal Studies Press, Canberra.

— 2009b, *The Gunditjmara Land Justice Story*, Native Title Research Unit, AIATSIS, Canberra.

— 2011, *Karajarri: A West Kimberley Experience in Managing Native Title*, Discussion Paper, Native Title Research Unit, AIATSIS, Canberra.

Worster, D 2008 [1997], *Nature's Economy: A History of Ecological Ideas*, 2nd edition, Cambridge University Press, Cambridge.

Yu, S 1999, *Ngapa Kunangkul: Living Water. Report on the Indigenous Cultural Values of Groundwater in the La Grange Sub-basin*, prepared for the Water and Rivers Commission, University of Western Australia, Perth.

Cases

Mabo v Queensland [No 2] (1992) 175 CLR 1

Onus v Alcoa of Australia Ltd (1981) 149 CLR 27

Western Australia v Ward [2002] 213 CLR 1 (8 August 2002)

Yanner v Eaton 1999 201 CLR 351

Yorta Yorta Aboriginal Community v the State of Victoria and Ors [1998 unreported]

Legislation

Fauna Conservation Act 1974 (Qld)

Native Title Act 1993 (Cth)

Wild Rivers Act 2005 (Qld)

Conservation and Land Management Act 1984 (WA)

2. Connections of Spirit: Kuninjku Attachments to Country

Luke Taylor

This place, the creek and water, we love this country, we Aboriginal people. We love it. The old people were the same, attached to the water and this land. The old people, our grandfathers and grandmother's, great grandparents, our ancestors, they lived here in this place, put here for them. That's how we talk about our land.

Our spirits lie in the water....

When we camp by the creek it soothes our spirits and keeps us cool. We understand it at places where my father took us, and my grandfather, mother's mother and father's mother. Today we want to continue to teach each other these things so we can understand. We did not invent this ourselves. The first ancestors from long ago are the origin. (Namirrkki, 2004: 112)

This is the way Ivan Namirrkki, born in 1961 and a Kuninjku language speaker from western Arnhem Land, describes his attachment to his country and its waters. Namirrkki is musing on his relationship to a camping place called Wakyoy on Manggabor Creek, a tributary of the Liverpool River, that is a short walk from his outstation at Kumurrulu. In a couple of sentences he has outlined an attachment which speaks to his understanding of creation and the intrinsic bodily link between himself and the place where he is living. The country was 'put there' by the Ancestral beings for Namirrkki's family and his continued maintenance of this country venerates these beings and the succession of human ancestors who have also lived there. There is a power in country that radiates to all those who live there and becomes incorporate in humans by virtue of their spiritual makeup. In this chapter I want to trace Kuninjku thinking about their connection to country to explicate Ivan's statement above. In doing so I wish to reveal the local conception of country that Kuninjku focus upon as important to their contemporary existence.

Kuninjku cast this relationship as a spiritual one, as a direct connection with Ancestral creator beings, and this religious outlook shapes the beliefs and values which are the framework for interpreting the meaning of their actions in the world (Taylor, 1996). I wish to focus on one man's understanding of his interrelations with the Ancestral powers, landscape features, and species of his country as an example of the broader Kuninjku perspective. In examining these connections we find explanations for Namirrkki's motivation to remain living on his country, particularly the importance of establishing the outstation at Kumurrulu. One context for this analysis is a recurring debate about the most appropriate mode of development for Aboriginal and Torres Strait Islander peoples living in remote regions.

The Kuninjku are traditional owners of lands to the south-west of the town of Maningrida in Arnhem Land in the Northern Territory (Altman, 1987; Taylor, 1996). All of Arnhem Land is formally owned by Aboriginal people as a result of the *Aboriginal Lands Rights Act Northern Territory 1976* (Cth). Today many Kuninjku families live on outstations or homeland centres on their own lands and regularly commute to Maningrida for services and supplies. Occasionally families will reside in the larger town in the wet season when transport to outstations may be difficult or for social reasons such as avoidance after a death or to ensure regular health care for a new born.

For some outsiders, these homelands appear to be an anachronism. For example Hughes (2007) points to the lack of development opportunities at such locales and criticises the effectiveness of the decentralised delivery of schooling and health services entailed by such living arrangements. The former Federal Minister for Aboriginal Affairs described these locations as 'cultural museums' (Vanstone, 2005). The current Northern Territory government policy does not support the funding of new outstations and otherwise promotes a reduction in government support of outstation infrastructure while favouring the support of 'growth towns' (Northern Territory Government, 2009). There is a belief that the government can create stronger employment opportunities in the towns and argument as to whether Aboriginal families should be compelled to move to larger settlements in order that services become more effective. Such measures that impact upon where Aboriginal people live can be considered a subset of a much broader program of intervention and enforced behavioural change associated with the Northern Territory Emergency Response implemented in June 2007.

The circumstances of outstation life vary across the continent and the example I describe here may not be typical. In respect to Kuninjku I argue that these more remote communities are not anachronisms so much as modern and culturally appropriate adaptations to the changed circumstances of their life. Other authors (such as Merlan, 1998, 2005 and Hinkson and Smith, 2005) have explicated how it is appropriate to consider the profoundly intercultural aspects of contemporary Aboriginal life. New expressions of Aboriginality are produced through intercultural experiences and, in the Kuninjku case, homelands are an expression of an extremely resilient religious outlook which venerates Ancestral presence in the land as well as considerations of the multiple servicing benefits that derive from the creation of relatively permanent small settlements in close proximity to a town. As Povinelli (1993) and Altman (1987) have shown, Aboriginal ideas regarding the importance of maintaining a presence on the land and of 'working' it – through the performance of ceremony, maintenance of sites, establishment of outstations, and continued hunting and gathering activities – continue to be central in the changed and intercultural circumstances of Aboriginal lifeworlds in the Northern Territory. It is important to acknowledge that Kuninjku developed a strong experience of town life at Maningrida in the 1960s but that they rejected this form of permanent settlement and the overt assimilation programs that they experienced there. Rather, many Kuninjku prefer to live on their country but regularly commute to the town as they need to – they make use of the services of the town on their own terms. Authors such as Tatz (1964), Rowse (1998), Altman (1987) and Folds (2001) have highlighted the way that assimilationist policy proscriptions have a history of failure in the Northern Territory partly because they did not account for the continuing strength of Aboriginal sociality and in particular the connections with country. Altman (2007a, 2007b) has pointed to the way that current government development agendas regarding Aboriginal employment, enterprise development, education, housing, health, and service delivery more broadly have a strongly coercive aspect. He notes that these policies do not engage with Aboriginal views regarding development that is appropriate for their families and on their lands. One of the points I wish to develop in this paper is the depth of Kuninjku concern to remain on their lands. Altman reveals that another weak point of current policies is the lack of appreciation of structural impediments to Aboriginal engagement with mainstream economic opportunity in some locales. Many communities are simply too far removed from markets and marketable resources.

Outstations comprise a relatively new and sedentary form of settlement as opposed to pre-contact forms of mobility across vast areas of land (Parliament of the Commonwealth of Australia, 1987). Kuninjku country is now crosscut with the new infrastructure of roads, airstrips, telecommunications, solar electricity, bores to provide water supplies, and buildings used to deliver health and education services. In the siting of homelands, Kuninjku such as Ivan Namirrkki have negotiated the need to sustain connections with particular country as well as take advantage of new opportunities provided by new technologies. Altman and Hinkson (2007a) have made the point that new technologies, such as the motor vehicle, have already profoundly transformed Kuninjku lives and allow for new expressions of identity. Kuninjku modernity is geographically more expansive than precontact forms when we consider the frequency of travel from outstations to the services and sociality offered at Maningrida. Similarly the art market has facilitated some Kuninjku to engage with audiences on a world stage (Altman and Hinkson, 2007a: 198–199). Kuninjku continue to be engaged in processes of social change but they do this in a way that involves assertions of a unique religious outlook and associated core values as *bininj* or Kuninjku people even as they engage with *balanda*, or non-Aboriginal people. Chief among their developing political concerns is the need for *balanda* to respect their attachments to country.

Namirrkki's outstation at Kumurrulu

Namirrkki lives at an outstation called Kumurrulu on elevated land amid a stringybark forest overlooking the valley of Manggabor Creek. This locale is on the edge of the Arnhem Land escarpment where the stone country meets the floodplain of the Liverpool River. In comparison with many other Australian landscapes this region is tremendously fertile in terms of the variety of foods that may be hunted and collected and Namirrkki's family exploits multiple ecological zones in the near vicinity. The floodplains in particular are breeding grounds for many species of waterbirds and saltwater and freshwater fish abound in the rivers. Bush foods are supplemented by store foods including staples such as flour and bread, tea, sugar and tinned foods purchased in Maningrida during regular visits or delivered by plane or boat during the wet season.

A short ten minute drive brings the family to the favourite hunting camp of Komnudd near Marrkolidjban outstation at the confluence of the freshwater and saltwater sections of Muralidbar Creek, a tributary of the Liverpool River. There is good shade at this place as well as the mounded remains of cooking fires that go back for many generations. There was a hollow log coffin placed in the trees here so that the spirits of these dead people could look out over the country that made them 'happy'. Substantial seasonal catches of barramundi are made here and there is an important site for the Barramundi Dreaming in this region. Other saltwater and freshwater fish species congregate in this locale. The camp is also adjacent to the major floodplain of the Liverpool River. This vast expanse is inundated in the wet season. When the waters recede Namirrkki's family can spear barramundi trapped in the shallow waters or, by stalking through the reeds with shotguns, may hunt the many species of bird life including magpie geese, whistle duck, brolga, and ibis. As the floodplain dries off more extensively, and Namirrkki's family burn the remaining reeds, large numbers of freshwater turtle may be dug from the drying mud by observing their breathing holes and probing with iron rods. Feral water buffalo and pig often travel down the river valley and range out across the floodplain and these introduced species are also tracked and shot to provide major supplies of fresh meat that are shared among multiple families.

As the dry season progresses, Namirrkki's family also camp for short periods on the white sand beaches of Manggabor Creek. Trees along the bank provide deep shade and Namirrkki's family camp out with mosquito nets strung from stakes driven into the sand. At this time Namirrkki will gradually burn off different areas as the spear grass dries out. He may also pay host to visiting families who come to fish for different species during the day. As the creek transforms into a series of separate billabongs people can fish with lines for barramundi, saratoga, catfish or smaller freshwater species. Occasionally saltwater crocodiles are speared for food in these billabongs. As the waters become more shallow, mud mussels, file snake and yabbies can be collected.

Also at this time, a short journey by four wheel drive accesses the tidal reaches of the Liverpool River. Here people can fish for large fish species such as shark, stingray, barramundi and catfish. Mangrove forests adjacent to the river hold colonies of flying fox which are hunted with shotguns. Wallabies are often hunted where they are sheltered in cooler jungle areas adjacent to the river.

In the wet season all these areas become inundated with dangerous floodwaters and transport becomes much more limited. However during this time the rock country and forest areas that surround Kumurrulu remain accessible and it is also the time when many fruiting tree species such as green plum, red apple and white apple can be harvested. Kangaroo and emu can be hunted in the rock country, and buffalo and pigs in the billabongs and springs of the higher reaches of the freshwater streams. Kumurrulu has access to an airstrip near Marrkolidjban outstation and boat landing on the Liverpool River for the purchase of supplies if the family chooses to stay in residence at this time. However, if Namirrkki does not have a working vehicle, or if his family miss the sociality of the town, he may relocate to stay with relatives in Maningrida during the height of the wet season.

Namirrkki uses Kumurrulu outstation as a relatively permanent base from which he can exploit a variety of ecological zones that include the rock country, savannah woodlands, freshwater streams, saltwater rivers, patches of jungle, mangroves, and floodplains. Hunting and collecting can involve short day trips from the outstation on foot or by car or short term relocation to camps situated closer to these resources. Altman (1984: 36; 1987) identifies a similar set of land systems and patterns of land use by people living at nearby Mumeka outstation. Access to food resources is an extremely important pragmatic reason for occupying this locale and this pattern of use of the land is emulated by many other Kuninjku families in approximately 15 other Kuninjku outstations (Altman, 1987, 2003). Hunting and gathering remains a central form of working Kuninjku country. Elsewhere in the top end of the Northern Territory, Povinelli (1993) has also made the point that such activity establishes and reiterates the links between people and country and that hunting and gathering is continuous with more religious connections to country. Povinelli documents the creation of connections between contemporary humans and Ancestral presence inside the earth by means of the 'sweat' of human toil that can be sensed by Ancestral spirits and through the use of language that is appropriate to communicate to the spirits of country.

Country of the spirit

Namirrkki has spoken of his love for country particularly the soothing qualities of living adjacent to its important waters. There is also a confidence and peace derived from living in one's heartland that flows to all activities conducted there. An understanding of the importance of country provides the context for more developed understanding of Kuninjku concepts of personhood, sociality, power, and health, as well as local constructions of other frameworks of human experience such as aesthetic experience.

Namirrkki says that one of the important reasons that he lives at Kumurrulu is because it allows him to protect the important site for the Leech Dreaming at Yibalaydjyigod and for the Maggot Dreaming at Yirolk. Kuninjku call such sites *djang* and say that they were created by the Ancestral beings also called *djang*. The use of this same term reveals the identity between country and Ancestral species that is at the core of Kuninjku thinking about country and ecological relationships. For Kuninjku the Ancestral beings are the source of original animating life force or *kun-ngudj* and this power may be sourced at *djang* sites. For example, I have briefly mentioned the relationship between the Barramundi *djang* and the prolific catches of barramundi in the Marrkolidjban area. At this locale in the past increase ceremonies were performed where ochres from the *djang* site were cast into the waters in order to release the fertilising powers of the Ancestral Barramundi.

In Namirrkki's eyes Leech and Maggot are major *djang*. Leech relates to a particular spring in the swampy section of Manggabor Creek. The Maggot *djang* at Yirolk is a waterhole that features a rock emerging from the water and surrounded by waterlilies. It is said that all the spirits of deceased landowners are inside a cave in the rock where it is covered by water. Kuninjku say that young people can't touch the water or the waterlilies and yet old and powerful people can drink at this place. Damaging the site would release an uncontrolled pestilence of these species into the world and anger the Rainbow Serpent, *Ngalyod*, that protects the site. Namirrkki says there is a tunnel made by the Maggot *djang* under the ground to another site on Bat Island near Maningrida in Myeri clan lands. In this respect the journey of the Ancestral being is considered to link Namirrkki to a wider polity of clans in the region. Namirrkki dances with the owners of this site in recognition of their joint relationship to a species that travelled through both their countries. Namirrkki occasionally paints this subject with lattices in the painting that show connections between different waterholes in the forms of the tunnels or creeks that link the different sites. Thus alliances between regional groups are also articulated on the basis of shared concern for this species. Keen (1994) has elaborated on this point in respect to Yolngu groups in north-east Arnhem Land; the presence of similar ecological zones and species in the countries of different clan groups can be used as the basis for alliances of shared responsibility in respect to particular sets of Ancestral beings. Keen (1994: 123–124) elaborates that the heterogeneity of ecologies within a clan's land provides the basis for relations with multiple other clans in the region of the same patrimoiety and links to the multiple ceremonies celebrating the different beings of that moiety.

Leech feeds on the blood of living animals and humans and has a body form like a small snake. Maggot is involved in processes of bodily decay through eating away the flesh of the dead and is also snake-like in its bodily form. In the Ancestral period, these beings are said to have been much larger and very dangerous. However these beings are also considered to be potentially dangerous and harmful to humans in the present if they were to multiply uncontrollably through damage to the site. In their fondness for eating flesh, their snake-like form, and their association with death and decay these species are linked to the conception of the Rainbow Serpent. As Morphy (2008) has shown for the Yolngu Mangalili clan in eastern Arnhem Land, themes associating Maggot with death and decay are also linked with purification and rebirth in the context of mortuary ceremonies. In Kuninjku thinking these sacred sites are also considered to be sources of fertilising power and, through the agency of *Ngalyod*, are linked to ideas regarding the recycling of human souls.

Kuninjku consider that the powers that *djang* invest in the land also animate all humans. In particular Kuninjku gain their *kunmalng* or sacred soul from these sites. Such spirits living in the deep waters at *djang* locales can make women conceive and, upon death, the spirits are said to return to these places. People also gain their personal names from places in their country and often also from their father's father thus highlighting the theme of patrilineal inheritance of these places (Evans, 2003: 68). One way of talking about powerful sites in the country is to use the term *kubolk murrng-rayek*, meaning literally 'country of strong bones'. Old men also have a measure of this power and can be described as developing 'strong bones' in the same way. In these respects the country, Ancestral species, and contemporary humans can be considered to be spiritually continuous (see Taylor, 1996). Namirrkki's desire to live near these sites is primarily an expression of his responsibility to tend to the Ancestral powers of the site and to protect the souls of deceased people of previous generations, and the, as yet, unborn souls of new people. For Namirrkki, living near these powerful sites gives him spiritual strength. He uses the language of 'strong bones' to describe the way his knowledge of country gives him power.

Kuninjku speak to the spirits of country and say that the spirits speak back. Visitors who do not know the site are introduced to it by the landowner who rubs them with his sweat and calls to the relevant spirits. Being rubbed with sweat ensures the spirits do not smell out the strangers. In addressing the spirit world, land owners must use the correct language and employ a special linguistic register called *kundangwok* which is clan specific (Garde, 2004: 110–111). Visitors must be identified by their place in the local kinship system and landowners plead with the spirits not to harm them. Strangers who wander unsupervised across country and do not understand its protocols can cause considerable anxiety among landowners and may inadvertently anger the spirits. For example, Ivan Namirrkki's father Peter Marralwanga acted to protect the Yirolk site in 1973 when he erected a barricade of tin drums across the road to protect his country after damage to the site by an ignorant road grading party (Cooke, 1981). It is also common during hunting exploits to hear land owners pleading with the spirits to help them to find game.

Waterholes such as Yirolk are often large and still bodies of water. The creation stories relating to such sites often involve the original Ancestral species angering the Rainbow Serpent *Ngalyod* in some way so that *Ngalyod* rises from inside the earth swallows the other being and draws them back down and this act effects the creation of the site. *Ngalyod* is seen as the original and most powerful being and all subsequent Ancestral beings came out of her body. The conceptualisation of *Ngalyod*, often described as the earth 'mother', involves a consideration of a ubiquitous life force existing inside the earth (Taylor, 1990). All sites and Ancestral species in the region are linked on the basis that, hidden inside the earth and in deep waters, *Ngalyod* is tending to them all.

Kuninjku say that in the wet season *Ngalyod* rises from these waters into the sky to make the monsoon rains. The rainbow seen in the rain is considered to be a manifestation of *Ngalyod's* spirit and the huge energies of the wet season storms are expressions of *Ngalyod's* power. The rain brought by *Ngalyod* rejuvenates the earth, makes the grass grow high, the sap run in the trees, and animals such as the birds and fish of the floodplains repopulate the world. The well being of Kuninjku is explicitly linked with this cycle as they hunt for game after the storms. The water cycle is intrinsic to Kuninjku thinking about *Ngalyod*. *Ngalyod* reaches from the earth and into the sky to vomit the wet season rains. These rains flood low lying areas and the waters gradually run away to the sea and to the permanent water sources that are *djang* as *Ngalyod* swallows these waters in the dry season. Kuninjku say that *Ngalyod* creates links between these waterholes by tunnelling under the ground and they indicate that *Ngalyod* can be responsible for creating springs that bubble with fresh water at some distance from flowing creeks. Water is perceived as a conduit of Ancestral power and a common way this is articulated is through the belief that the spirits of unborn humans live at *djang* sites as small fish with a rainbow sheen.

One purpose of the recreation of Ancestral events in major regional ceremonies such as Kunabibi and Yabbadurruwa, is to ensure the arrival of the wet season and thus the release of the powers of *djang* through the agency of *Ngalyod*. Many participants from the multiple clans of the region gather for these ceremonies and the performance is said to ensure the good health of the participants and maintain the link between the multiplication of species and continued opportunities for good hunting. Specifically participants say that the ceremony makes them feels strong, healthy and 'fat'. The term 'fat' also has many other reverberations with health and vitality in the natural world; the ceremony makes the 'fat' or sap run in the trees and causes animals to put on the fat that is so highly valued. Elsewhere, McDonald (2003) has shown there are clear health benefits from eating the fat of bush animals as opposed to shop purchased meat. Other researchers have measured a correlation between outstation residence and improved health outcomes in this locale (Burgess *et al.*, 2005) thus lending objective support to Kuninjku perceptions of the link between living on country and achieving better health.

Following the ancestors

While Kumurrulu is a relatively new outstation established just a few kilometres from the larger Marrkolidjban outstation, Namirrkki's family have a long history of living in this area. Namirrkki explains that it was in discussion with his father Peter Marralwanga before his death in 1987 that they together decided that Kumurrulu would be a good location for a new outstation.

Marralwanga has described that he was born here and that he moved to live in different camp sites across the region throughout his life (see, Taylor, 1996: 65–69). Other Kuninjku say that Marralwanga received promised wives from families who lived in the Marrkolidjban region and that in recognition of a lifetime of association he became the acknowledged ceremonial leader and 'boss' for the locale. Government patrol officer Gordon Sweeney found many families living in this district during a patrol in 1939 (Sweeney, 1939). Another officer Syd Kyle-Little (1957: 215–216) reports many people living at Marrkolidjban in 1949 and Peter Marralwanga remembered him from this time. Patrol officer Ted Evans (1963) led an expedition to establish the Maningrida/Oenpelli road and found Kuninjku families camped nearby at the Liverpool River crossing in 1963. It is important to acknowledge that Marralwanga's own Kardbam patri-clan territory is further to the south-east. However Kuninjku explain that movement beyond one's own country was a characteristic of pre-settlement life as were negotiations to gain residence and ceremonial rights in new lands. Hiatt (1965: 18–20) reports similar patterns among the neighbouring Gidjingali.

Maningrida was established as a government settlement in 1957. However Peter Marralwanga was not generally interested in the attractions of permanent living in the town. This kind of existence was foisted upon his family in the 1960s as a feature of the development of Maningrida as a government run showcase of the effectiveness of assimilation policies. Kettle (1967: 206) notes that Marralwanga was eventually encouraged to move to live at Maningrida in 1963 to seek medical assistance for Namirrkki's elder brother who was then very young and showing early signs of leprosy. At Maningrida Kuninjku were required to eat in communal dining rooms and were trained to work in forestry, agriculture and fishing enterprises. Substantial infrastructure was established in the town to support these industries. Today Maningrida is surrounded by a grid of dirt tracks built to service tracts of cypress pine conserved for the forestry enterprise. An abandoned fire tower stands at the entrance to the town although, ironically, the forestry workers could never hope to control the multiple fires that swept through the stands of timber. Fires have been lit to manage lands in this region for millennia. Haynes (1978) documents the growing resentment of landowners to the expanding forestry activities at Maningrida and in particular a turning point in relations caused by damage to an important sacred site in 1971 by forestry workers bulldozing a road. While the Aboriginal men were encouraged to work in these enterprises, the women were trained as domestic servants and cleaned the homes of the white staff during the day. Haynes (1978) notes the anger about so many *balanda* working in the town and growing calls for removal of the forestry from the late 1960s and for Aboriginal self management at Maningrida in the early 1970s.

Over time the enterprises in the town gradually collapsed partly because of distance from markets, transport difficulties, and lack of motivation among Aboriginal people for the highly controlled lifestyle (Altman, 1987: 4–5, 18–19; Taylor, 1996: 36–37). In order to develop Maningrida some 13 different language groups from the north central Arnhem Land region were forced to live in close proximity and violent disputes were common. Peter Marralwanga, for example, carried the scar of spear wound in his chest received in 1964 during a major confrontation. He is reputed to have killed the other protagonist in this exchange.

In the early 1970s the changed policy of self-determination brought in by the Whitlam Government and the hearings of the Woodward Commission into land rights saw a dismantling of this form of enterprise development at Maningrida and encouragement for the different groups to return to their lands. Kuninjku were quick to return and establish outstations in their own lands. With logistical assistance from *balanda* staff including John Hunter the superintendent at Maningrida, Marralwanga became one of the senior founders of an early outstation at Marrkolidjban in 1973. Marrkolidjban outstation grew over the years and came to comprise a substantial assembly of buildings housing multiple family groups. Marralwanga had six wives in total in his life and numerous children. Marralwanga, Namirrkki, and other residents developed a reputation for the production of excellent bark paintings and women produced excellent basketry (Taylor, 1996). Marralwanga enforced a ban on the consumption of alcohol in the outstation to improve its social conditions and to check the visits by countrymen who occasionally brought alcohol from Oenpelli to Maningrida and passed close to the outstation.

While many families still choose to live in the town, Maningrida today is a service centre for some 800 people who live in a total of 32 outstations in an area of 10,000 square kilometres. Kuninjku homelands consist of mudbrick houses, provision for communal facilities such as an outstation school or women's centre, toilets and showers, an airstrip, road access and occasionally a local boat landing, a solar power water bore, and solar powered telephone. As Altman and Hinkson (2007a) report, the tension between permanently sited infrastructure and the need to access country is partly negotiated through the hyper-mobility afforded by motor transport. Kuninjku motor vehicles are rarely idle. Outstation facilities are maintained through the Bawinanga Aboriginal Corporation established in Maningrida to service outstation residents. Schooling is provided through a specifically designed outstation school program attached to the Maningrida School. *Balanda* teachers visit the outstations for a number of days each week and work with residents of the outstation employed as teacher aids. As detailed above, there is a tension involved in maintaining this new and relatively permanent infrastructure and the desire of most outstation residents to establish less permanent camps in order to exploit seasonal food abundances, visit relatively remote *djang* sites, or to engage in social activities such as the performance of ceremonies. However service deliverers based in Maningrida have developed a measure of expertise around Kuninjku mobility and, through good communications, can generally manage to accommodate this shifting pattern of residence. The Bawinanga Aboriginal Corporation has used the Australian Government funded Community Development Employment Program in a manner that recognised the work involved in tending to country, managing outstations, developing infrastructure such as housing and roads, as well as for the development of community enterprises such as Maningrida Arts and Culture, a shop and mechanical workshop for outstation residents. However the federal government has moved to change the Community Development Employment Program from its use for part time employment support in remote Indigenous communities to a scheme to ready participants to find full time work outside of the scheme. Former Community Development Employment Program participants will now complete work training activities and receive income support payments until they can find full time employment. Government has funded service delivery positions and some ranger positions to assist in the creation of full time work but the number is by no means commensurate with the need in Maningrida.

With the death of his father, Ivan Namirrkki established his outstation at Kumurrulu adjacent to Marrkolidjban. At present the infrastructure at Kumurrulu consists of only one house for Namirrkki and his sons as well as bore and telephone. Namirrkki exhibits strong emotion when he describes how his father guided him to make an outstation near this place. Kumurrulu is an offshoot of the larger Marrkolidjban outstation a few kilometres down the road and now occupied by his elder brother Bill Birriyabirriya. Kumurrulu is one of a second wave of outstations developed as the outstations established in the 1970s become too large. Larger outstations can develop social tensions and fission of large communities is a characteristic way that Kuninjku address such problems. The smaller outstation provides peace and quiet for Namirrkki's family away from Marrkolidjban and yet it is only a short drive to the school located there.

It is important to acknowledge that Namirrkki's outstation is situated in Barbinj clan land although his own Kardban clan country is further south near Mankorlod. Like his father before him he is a recognised boss of these new lands. Kuninjku still strongly recognise the mosaic of patrilineal clan land ownership in western Arnhem Land although they also acknowledge that historical circumstance has seen considerable movement of individuals outside of their own clan territories. Namirrkki's rights in this country have been negotiated over time primarily through his continued residence and intimate knowledge of this locale. This does not mean that the clan associations of country have been blurred, rather they are acknowledged in the present as are the personal historical circumstances that have led to Namirrkki's intimate knowledge of this country. In many respects Namirrkki benefits from the powerful status of his father although it is also true the Namirrkki himself has been politically active in his own lifetime in tending to these lands. Caring for important ceremonial sites is a key activity in this context and for Kuninjku this constitutes extremely valuable work. Namirrkki's status is respected by the jural public of other landowners who jointly maintain Kuninjku ceremonial life. There is a major site for performance of the Kunabibi ceremony near Marrkolidjban and Namirrkki is now broadly recognised as a major *djungkay* or ceremonial manager for performances of the ceremony at this place. Namirrkki was recently one of the senior hosts for a visit by hundreds of people from the broad region for a performance of this ceremony near his home in 2007.

Country based enterprises

My concern has been to explicate some components of the way that contemporary Kuninjku interact with their country and their interpretations of this coexistence. The fundamental bedrock of this connection is Kuninjku belief about the enduring Ancestral energies in their country and their manifestation as life force and human 'spirit'. The important point is that this religious perspective is still central to contemporary Kuninjku life choices. Adaptations to new forms of settlement of their land, new kinds of work effort, and development of new economic opportunities all articulate a concern to nurture this life force at the same time as articulating connections with western or *balanda* modes of action and thinking. As Merlan (2005) has shown, Aboriginal discourses can serve to reconstruct Aboriginal identity as distinct from western identity in contemporary intercultural contexts. Kuninjku articulate an understanding of their distinctiveness even as they actively engage with these new opportunities.

Kuninjku have readily engaged in new forms of enterprise development suited to their relative remoteness from markets. The production of art and craft was facilitated from the earliest days of settlement of Maningrida and was particularly encouraged to support outstations (Altman, 2004; Taylor, 1996). Recently Kuninjku living at outstations have engaged with more formalised ecological management and cultural tourism programs. Altman (2003, 2005, 2007a) has outlined how Kuninjku can meld an extremely strong desire to maintain an economy of food gathering with income support provided by governments and opportunities to work in market based enterprises to develop what he calls a 'hybrid' economy at Kuninjku outstations. Elements of this concept have been incorporated in the Australian Conservation Foundation's promotion of the 'culture and conservation economy' for Aboriginal communities in northern Australia (Hill *et al.*, 2008: 4–5, 19–21). These characterisations recognise the increasing preference, in some remote Aboriginal lands, for the development of sustainable country based enterprise.

Art market

Just as the development of outstations can be seen as the negotiation of a new form of expression of Kuninjku identity, Kuninjku participation in the world art market is valued by Kuninjku because it allows for 'new' expressions of identity on a global platform. The market for bark paintings is a very important way of earning an income and an opportunity to teach *balanda* about Kuninjku culture. Painting is a key means of communicating knowledge about the spiritual power of Kuninjku lands both among Kuninjku and to broader world audiences (Taylor, 1996, 2008).

Namirrkki often collects bark for painting from the large stands of *Eucalyptus tetradonta* or stringybark that surround his outstation. Red and yellow ochres can be collected from river and creek crossings. Sacred ochres can also be collected at particular locales along Marrkolidjban Creek. Namirrkki says that his spirit and culture is in the ground at *djang* sites and when he paints he brings it 'out', brings it forth for all to see (Namirrkki, 2004: 112). Namirrkki was taught many of the stories about his lands from his father as a feature of learning to paint. Often Marralwanga and Namirrkki used to work jointly on a single painting in order that the skills and knowledge could be transferred as part of the younger artist's apprenticeship. On one occasion in 1983 Namirrkki travelled with his father to Perth to attend an exhibition of joint work and this stimulated him to take on a stronger role as a professional artist. Later in his life Namirrkki travelled more broadly in Australia for both group and individual exhibitions (see, Perkins, 2004: 220). Namirrkki can participate in the world art market through the support provided by Maningrida Arts and Culture, a Bawinanga Aboriginal Corporation enterprise, based in the town. Importantly his inspiration comes from sacred places very close to his home including the sites for Leech and Maggot described above. He also paints important *djang* such as Goanna, *Ngalyod* and the *Nakorrkko* creator beings from the lands of his Kardbam patri-clan. There is a strong sense of wellbeing derived from this repeated engagement with imagery of his country and a sense of spiritual connection with previous generations of Kuninjku, particularly his father, also linked with these lands.

Kuninjku aesthetics is grounded in religious conviction. The dazzling designs of Kuninjku paintings evoke the powers that radiate from *djang* sites. A common subject painted by Peter Marralwanga and Namirrkki was *Ngalyod* the Rainbow Serpent. The figure was often shown as a great twisting snake that surged with life force. Dazzling designs on the figure's body were suggestive of its rainbow manifestation. The point of painting such work for the market is to expose viewers directly to the power of the Ancestral realm. Following a similar line of argument to that of Deger (2006) in relation to Yolngu use of new film media and Tamisari (2005) in relation to Yolngu dance performance, I argue (Taylor, 2008) that Kuninjku engage with the new opportunities afforded by the art market in order that the aesthetic force of their paintings 'opens' *balanda* viewers to a new way of seeing the world which has Ancestral presence at its centre.

Feelings for country also inspire innovations in the market context. The style of painting from this region has undergone a major transformation in the last 20 years (Taylor, 2008). Kuninjku artists have moved away from depicting figurative subjects to depicting key sacred waterholes in a more geometric ceremonial style called Mardayin. This style of painting relates to ceremonial body paintings of the same name. Namirrkki for example paints the sites Leech and Maggot and other nearby sites in this geometric format (Perkins, 2004: 86). Kuninjku have perceived a greater importance of revealing their ceremonial knowledge about country in this new format and there has been a positive response by the market for these more important paintings. Kuninjku seem more self-conscious about land as spiritual heartland needing to be protected and painting is perceived in a more political sense as one way of continually reasserting the importance of country and its attached values on a world stage.

Ecological services

Namirrkki's knowledge of the land also provides other economic advantages for his family. Namirrkki and his son Obed work in the Djelk Rangers established by the Bawinanga Aboriginal Corporation and based in Maningrida. The ranger program was established in 1991 as a feature of the Community Development Employment Project and is maintained now through the Working on Country and the Community Development Employment Project and partnerships with multiple other state and federal agencies. Intimate knowledge of country allows participants to engage in land management activities such as feral animal and weed control, species survey and monitoring programs, controlled burning, detection of incursions by foreign fishing vessels and participation in new enterprises such as the breeding and sale of crocodile and long necked turtle hatchlings and recently tarantula spiders (see Fordham *et al.*, 2010a, b). Rangers have also helped botanists, zoologists, ecologists, linguists, anthropologists, and rock art researchers in numerous joint projects in recent years (see, for example, Bowman and Robinson, 2002; Garde, 2004; Telfer and Garde, 2006).

As a senior landowner Namirrkki's status is often revealed in his skill in the appropriate burning-off of country as the dry season advances and there is now a considerable body of research data on the importance of this activity in this region (Haynes, 1985; Bowman *et al.*, 2001; Yibarbuk *et al.*, 2002; Russell-Smith *et al.*, 2009). One of co-workers for Bowman, Garde and Saulwick's (2001) research was Big Bill Birriyabirriya, Ivan Namirrkki's elder brother, living at Marrkolidjban. Through detailed interviews conducted in the appropriate language, these authors were able to ascertain the way that burning is used in the hunting of kangaroos through promoting the regrowth upon which the animals feed and also through the highly controlled and communal activity of fire drives. This research also shows the importance, to local landowners, of burning early in the dry season so as not to damage flowering fruit and honey trees and to avoid the development of uncontrollable fires that burn too hot and might kill trees. Recently Garde and others have added to this picture by presenting extensive translated conversations about ecological relationships and the use of fire in the rock country region adjacent to Kuninjku lands (Garde *et al.*, 2009). Senior landowners have much to teach younger people employed as rangers in this regard.

Rangers from the Djelk group collaborate with neighbouring ranger groups participating in the West Arnhem Land Fire Abatement Project (Whitehead *et al.*, 2009). As participation in this project shows, Aboriginal habitation of western Arnhem Land and participation in low intensity burning practices involve reduced emissions of greenhouse gases when compared to uncontrolled wildfires in depopulated regions. The fire management activities of the Aboriginal people living in this area are enhanced through the use of region wide satellite tracking and planning of fires as well as aerial incendiary burning of areas of the stone country that are otherwise hard to access. These researchers argue in support of the beneficial environmental effects provided through people living on outstations and making use of these new technologies and for the potential of new income streams from reducing carbon emissions as an innovative model of economic engagement for outstation residents.

The Djelk group and traditional owners in the region have recently negotiated a new Indigenous Protected Area for their country which will see the more permanent establishment of ecological services as contemporary employment on their homelands (Australian Government, 2010). The Djelk Indigenous Protected Area includes over 673,200 hectares of land that has been voluntarily provided for conservation of natural species by the traditional owners. Recognition of an IPA requires management of lands by the owners to internationally agreed standards for sustainable use and habitat protection. The traditional owners can draw upon multiple government and non-government programs, such as Working on Country, that support employment for the management of such lands (see also, Bauman and Smyth, 2007). Rangers are currently being moved from employment under the Community Development Employment Project to these other support programs, although the number of positions that can be supported this way is very small. It must also be acknowledged that this work is highly collaborative between Aboriginal people and *balanda* and requires the development of sustainable social relations between landowners, government departments, and land management organisations if the work is to be successful (Whitehead *et al.*, 2009: 295–302; see also Hill *et al.*, 2008 for other locations). Certainly the particular religious outlook and skills of Kuninjku provide an advantage in respect to their engagement in such activity (Altman 2003, 2005, 2007a; Burgess *et al.*, 2005).

There are also incipient developments in cultural tourism to visit rock art sites and homeland centres on Kuninjku lands through Bawinanga Aboriginal Corporation Tourism. Kuninjku landowners manage the tour to ensure that the movements of tourists are controlled. Tourists are now seeking to move beyond the relatively developed national park areas such as Kakadu, outside of Arnhem Land to the west, and are seeking stronger cultural engagements and direct access to Aboriginal perspectives.

As Altman has revealed, it is the comparatively late impact of colonisation on Kuninjku lands that has given them an advantage in these new enterprise developments. If Kuninjku lands had not been relatively spared from more intrusive forms of development they would not experience the advantages that are now being presented (Altman, 2007a: 4–7; see also, Bauman and Smyth, 2007).

It is important to realise that the engagements with Australian scientists and land managers that this work entails is new activity for Kuninjku. Using GPS to track the spread of introduced animal and plant species, application of poisons, helicopter and quad bike surveys and culls, management of tourists, rock art recording, husbandry of particular species for the aquarium market, and aerial burning and satellite tracking of burn scars in region wide planning and management of carbon emissions are all skills that have been grafted to existing knowledge and management of country and species. Kuninjku ideas regarding the ceremonial management of country are not identical to land management and conservation in the sense that environmental agencies espouse (see also, White and Meehan, 1988: 37–39; Horton, 2000: 127–140; Fordham *et al.*, 2010a). Kuninjku concern for species, people and country is embedded in broader considerations of the maintenance of Ancestrally instituted conventions of behaviour that ensure the preservation of Ancestral spirit and power through time. Indeed it took many years of development training and ongoing support by staff of the Bawinanga Aboriginal Corporation, development of specific formal training programs, as well as engagements with and the Caring for Country Unit of the Northern Land Council to establish understanding of the elements of the position of 'ranger' and to inculcate the regimen of this new kind of work among Kuninjku (Fordham *et al.*, 2010a). The Maningrida School now has a junior ranger program for senior students (Schwab, 2006).

The establishment of the Indigenous Protected Area requires recognition by local people of externally developed standards for maintaining country and species in order to negotiate access to the resources and jobs to conduct these more formalised management activities. It might be expected that managing these different perspectives in the coming years will require considerable negotiation. Nevertheless Kuninjku are now very supportive of the 'two-way' methods involved in the management of their country and there appear to be developing convergences in thinking about country at a deeper level. In these intercultural developments, Kuninjku are learning the scientific framework of western technologies of ecology and conservation and also teaching *balanda* much about the characteristics of different species and the particular ecological interactions in this unique landscape. These are new kinds of activity and should not be simply subsumed as a continuation of pre-existing Kuninjku practice so much as developments that Kuninjku consider appropriate to the way they wish to engage with country, and with *balanda*, in the present.

It is wrong to identify the religious outlook of people such as Kuninjku as some sort of barrier to be overcome in these developments. Rather, Kuninjku religion provides a base that supports their grasp of more contemporary ecological thinking. As authors such as Latour (1991) and Rose (2005) and Weir (2009) in the Australian Aboriginal context, have pointed out, theories involving hyper-separations of humanity from nature promote a view of nature as a resource and an overconfident attitude in respect to the human technological/ scientific capacity to manage our environment. Such views stem from the Judeo-Christian/Stoic theological root and were exacerbated following the Enlightenment (Bennett, 1983; Latour, 1993; Weir, 2009). Understanding of the ways that ecological destruction is ultimately resulting in damage to humanity requires an appreciation of the complex webs of connectivity of humans in nature. Comprehension of the impacts of ecological destruction involves not just the development of new branches of science focused on ecological complexity and locational particularity but also critical review of fundamental theoretical tenets of western science and theology. In as much as the Kuninjku religious outlook highlights a perspective of connectedness, and their engagement with the natural world requires detailed observation, there is much that they share with the contemporary ecologists with which they work (see, Rose, 2005; Fordham *et al.*, 2010a). In the Maningrida context we might contrast the hubris of the attempt to create a forestry industry in the region in the 1960s with the current investigation of employment in ecologically informed and sustainable conservation activities.

Conclusion

The chapter opened with a quote from Ivan Namirrkki who was elaborating on his feelings for country. The quote was obtained in a discussion with Namirrkki about why he engages with the world art market (Namirrkki, 2004). The quote involves an assertion that the powers in the earth are not 'invented' but true and necessarily need to be acknowledged by outsiders as well as Kuninjku.

Rather than addressing the topic of ecology as it is more broadly understood, this chapter has focused upon elaborating Kuninjku thinking about country and the continuing powers of *djang*. This conceptualisation reveals the way that Kuninjku link ideas about human identity to places and to species via an appreciation of their animating Ancestral spirit. Human relationships and political constructions, human responsibility in respect to certain places, human interactions with other species, relations between non-human species, considerations of major processes of seasonality and change in the natural world, the aesthetics of art and performance, and emotions and feeling for country are all interlinked on this basis. Similarly new activities enabled as a feature of the changed circumstances of Arnhem Land life are now related into this outlook that has spiritual connectivity at its heart.

There is a growing self-consciousness among Kuninjku of the need to make *balanda* understand these things and of the threats posed by those who do not. Indeed many of the enterprise engagements of Kuninjku such as the art market or environmental services involve communication with *balanda* about the value of the Ancestral world for political as well as economic reasons. Although Kuninjku are willing partners in the two way exchanges that characterise their new circumstance this does not mean that they necessarily accept the broader range of *balanda* values. Through their engagements in these new activities Kuninjku also readily express their differences from *balanda*.

The homelands movement was not the dream of left wing policy ideologues to preserve a passing world so much as an accommodation by *balanda* of Aboriginal agency in creating new forms of social development where country, understood as a spiritual unity of people with the life of a place, was a central concern. This new form of settlement involved an explicit rejection of attempts at assimilation in the 1960s and economic development as practiced at Maningrida at that time. Kuninjku voted with their feet in the early 1970s by moving back to country even though support was very limited in the early years. Rowse (1998) too has shown how the under-appreciation of Aboriginal attachments to country frustrated work force proposals and assimilationist policies in central Australia during this period. Folds (2001) reveals in detail the means by which the Pintupi from central areas of the Northern Territory maintained their concern to preserve local cultural values and undermined and transformed successive mainstream programs aimed at assimilation. Aboriginal people have the relative power to effect these transformations in remote situations and to confound policies and programs issued by central bureaucracies and governments. The more successful alternative involves respect for the creative way that Aboriginal people are melding distinct knowledge and value frameworks in these locales and the development of programs that seek Aboriginal engagement on these terms.

Kuninjku have already identified appropriate forms of development in their country that assert core cultural values in new activities. Mainstream or coercive development proscriptions that ignore Kuninjku belief and attempt to divorce Kuninjku from their country, and impose conceptions of 'real' work, 'real' education, or 'real' health options will fail in the same way as earlier attempts at assimilation because they cannot engender Aboriginal engagement.

Considerations of development must come to grips with the fundamental epistemological challenge of Kuninjku belief of Ancestral creation – a belief from which all other conceptions of reality flow.

References

Altman, J 1984, 'The dietary utilisation of flora and fauna by contemporary hunter-gatherers at Momega outstation, north-central Arnhem Land', *Australian Aboriginal Studies* 1984(1): 35–46.

— 1987, *Hunter Gatherers Today: An Aboriginal Economy in North Australia*, Australian Institute of Aboriginal Studies, Canberra.

— 2003, 'People on country, healthy landscapes and sustainable economic futures: the Arnhem Land case', *The Drawing Board: An Australian Review of Public Affairs* 4(2): 65–82.

— 2004, 'Brokering Kuninjku art: artists, institutions and the market', in *Crossing Country; The Alchemy of Western Arnhem Land Art*, H Perkins (ed), Art Gallery of New South Wales, Sydney: 173–187.

— 2005, 'Development options on Aboriginal land: sustainable Indigenous hybrid economies in the twenty-first century', in *The Power of Knowledge: The Resonance of Tradition*, L Taylor, GK Ward, G Henderson, R Davis and LA Wallis (eds), Aboriginal Studies Press, Canberra: 34–48.

— 2007a, 'Alleviating poverty in remote Australia: the role of the hybrid economy', *CAEPR Topical Issue No. 10/2007*, Centre for Aboriginal Economic Policy Research (CAEPR), The Australian National University (ANU), Canberra.

— 2007b, 'In the name of the market?', in *Coercive Reconciliation: Stabilise, Normalise, Exit Aboriginal Australia*, J Altman and M Hinkson (eds), Arena Publications Association, North Carlton, Victoria: 307–321.

— and M Hinkson 2007a, 'Mobility and modernity in Arnhem Land: the social use of Kuninjku trucks', *Journal of Material Culture* 12(2): 181–203.

— and M Hinkson (eds) 2007b, *Coercive Reconciliation: Stabilise, Normalise, Exit Aboriginal Australia*, Arena Publications Association, North Carlton, Victoria.

Bauman, T and D Smyth 2007, *Indigenous Partnerships in Protected Area Management: Three Case Studies*, Australian Institute of Aboriginal and Torres Strait Islander Studies, Canberra.

Bennett, DH 1983, 'Some aspects of Aboriginal and non-Aboriginal notions of responsibility to non-human animals', *Australian Aboriginal Studies* 1983(2): 19–24.

Bowman, DMJS, M Garde and A Saulwick 2001, 'Kunj-ken makka man-wurrk: fire is for kangaroos: interpreting Aboriginal accounts of landscape burning in central Arnhem Land', in *Histories of Old Ages: Essays in Honour of Rhys Jones*, A Anderson, I Lilley and S O'Connor (eds), Pandanus Books, Canberra: 61–78.

Bowman, DMJS and CJ Robinson 2002, 'The getting of nganabbarru: observations and reflections on Aboriginal buffalo hunting in northern Australia', *Australian Geographer* 33(2): 191–206.

Burgess, CP, FH Johnston, DMJS Bowman and PJ Whitehead 2005, 'Healthy country: healthy people? Exploring the health benefits of Indigenous natural resource management', *Australian and New Zealand Journal of Public Health* 29(2): 117–122.

Cooke, Peter 1981, 'An exhibition of paintings by Djakku (Peter Marralwanga)', Maningrida Arts and Crafts, Maningrida.

Deger, J 2006, *Shimmering Screens: Making Media in an Aboriginal Community*, University of Minnesota Press, Minneapolis, Minnesota.

Department of Environment, Water, Heritage and the Arts (DEWHA) 2010, *Djelk Arnhem Coast, Northern Territory*, Department of Environment, Water, Heritage and the Arts, Canberra, available at: <http://www.environment.g.au/indigenous/ipa/declared/djelk.html>

DEWHA – *see* Department of Environment, Water, Heritage and the Arts

Evans, N 2003, *Bininj gun-wok: A Pan-dialectical Grammar of Mayali, Kunwinjku and Kune*, Pacific Linguistics, The Australian National University, Canberra.

Folds, R 2001, *Crossed Purposes: The Pintupi and Australia's Indigenous Policy*, University of New South Wales Press, Sydney.

Fordham, A, W Fogarty, B Corey and D Fordham 2010a, *Knowledge Foundations for the Development of Sustainable Wildlife Enterprises in Remote Indigenous Communities of Australia*, Centre for Aboriginal Economic Policy Research, ANU, Canberra.

— 2010b, *The Viability of Wildlife Enterprises in Remote Indigenous Communities of Australia: A Case Study*, Centre for Aboriginal Economic Policy Research, ANU, Canberra.

Garde, M 2004, 'Growing up in a painted landscape', in *Crossing Country; The Alchemy of Western Arnhem Land Art,* H Perkins (ed), Art Gallery of New South Wales, Sydney: 107–111.

— in collaboration with Bardayal Lofty Nadjamerrek, Mary Kolkkiwarra, Jimmy Kalarriya, Jack Djandjomerr, Bill Birriyabirriya, Ruby Bilindja, Mick Kubarkku and Peter Biless 2009, 'The language of fire: seasonality, resources and landscape burning of the Arnhem Land plateau', in *Culture, Ecology and Economy of Fire Management in North Australian Savannas Rekindling the Wurrk Tradition,* J Russell-Smith, PJ Whitehead and P Cooke (eds), CSIRO Publishing, Collingwood, Victoria: 85–164.

Haynes, CD 1978, 'Land, trees and man (*gunret, gundulk, dja bining*)', *Commonwealth Forestry Review* 57(2): 99–106.

— 1985, 'The pattern and ecology of *munwag*: traditional Aboriginal fire regimes in north-central Arnhemland', *Proceedings of the Ecological Society of Australia* 13: 203–214.

Hiatt, L 1965, *Kinship and Conflict: A Study of an Aboriginal Community in Northern Arnhem Land,* The Australian National University, Canberra.

Hill, R, EK Harding, D Edwards, J O'Dempsey, D Hill, A Martin and S McIntyre-Tamwoy 2008, *A Cultural and Conservation Economy for Northern Australia,* Land and Water Australia, Canberra.

Hinkson, M and B Smith 2005, 'Introduction: conceptual moves towards an intercultural analysis', *Oceania* 75(3): 157–167.

Horton, D, 2000, *The Pure State of Nature: Sacred Cows, Destructive Myths and the Environment,* Allen & Unwin, St Leonards, New South Wales.

Hughes, H 2007, *Lands of Shame: Aboriginal and Torres Strait Islander 'Homelands' in Transition,* Centre for Independent Studies Ltd, St Leonards, New South Wales.

Keen, I 1994, *Knowledge and Secrecy in an Aboriginal Religion,* Clarendon Press, Oxford.

Kettle, E 1967, *Gone Bush,* The Devonshire Press, Sydney.

Kyle-Little, S 1957, *Whispering Wind: Adventures in Arnhem Land,* Hutchinson & Co, London.

Latour, B 1993, *We Have Never Been Modern,* Harvard University Press, Cambridge, Mass.

McDonald, H 2003, 'The fats of life', *Australian Aboriginal Studies* 2003(2): 53–61.

Merlan, F 1998, *Caging the Rainbow: Places, Politics, and Aborigines in a North Australian Town*, University of Hawai'i Press, Honolulu.

—— 2005, 'Explorations towards intercultural accounts of socio-cultural reproduction and change', *Oceania* 75(3): 167–183.

Morphy, H 2008, '"Joyous maggots": the symbolism of Yolngu mortuary rituals', in *An Appreciation of Difference: WEH Stanner and Aboriginal Australia,* M Hinkson and J Beckett (eds), Aboriginal Studies Press, Canberra: 137–150.

Namirrki, I 2004, 'Our spirits lie in the water', in *Crossing Country; The Alchemy of Western Arnhem Land Art,* H Perkins (ed), Art Gallery of New South Wales, Sydney: 112–113.

Northern Territory Government 2009, 'Working future: Outstations/homelands policy', Department of Chief Minister, Northern Territory Government, Darwin, available at: <http://www.workingfuture.nt.gov.au/download/ Headline_Policy_Statement.pdf>

Parliament of the Commonwealth of Australia 1987, *Return to Country: The Aboriginal Homelands Movement in Australia*, House of Representatives Standing Committee on Aboriginal Affairs, Canberra.

Perkins, H (ed) 2004, *Crossing Country; The Alchemy of Western Arnhem Land Art,* Art Gallery of New South Wales, Sydney.

Povinelli, EA 1993, *Labor's Lot: The Power, History, and Culture of Aboriginal Action,* The University of Chicago Press, Chicago.

Rose, D 2005, 'An Indigenous philosophical ecology: situating the human', *Australian Journal of Anthropology* 16(3): 294–305.

Rowse, T 1998, *White Flour, White Power: From Rations to Citizenship in Central Australia*, Cambridge University Press, Cambridge.

Russell-Smith, J, PJ Whitehead and P Cooke (eds) 2009, *Culture, Ecology and Economy of Fire Management in North Australian Savannas Rekindling the Wurrk Tradition,* CSIRO Publishing, Collingwood, Victoria.

Schwab, RG 2006, *Kids, Skidoos and Caribou: The Junior Canadian Ranger Program as a Model for Re-engaging Indigenous Australian Youth in Remote Areas,* Centre for Aboriginal Economic Policy Research, ANU, Canberra.

Tamisari, F 2005, 'The responsibility of performance; the interweaving of politics and aesthetics in intercultural contexts', *Visual Anthropology Review* 21(1&2): 47–62.

Tatz, CM 1964, 'Aboriginal administration in the Northern Territory of Australia', unpublished PhD thesis, Australian National University, Canberra.

Taylor, L 1990, 'The rainbow serpent as visual metaphor in western Arnhem Land', *Oceania* 60: 329–344.

— 1996, *Seeing the Inside: Bark Painting in Western Arnhem Land*, Clarendon Press, Oxford.

— 2008, '"They may say tourist, may say truly painting": aesthetic evaluation and meaning of bark paintings in western Arnhem Land, northern Australia', *Journal of the Royal Anthropological Institute (NS)* 14: 865–885.

Telfer, WR and MJ Garde 2006, 'Indigenous knowledge of rock kangaroo ecology in western Arnhem Land, Australia', *Human Ecology* 34(3): 379–406.

Vanstone, A 2005, 'Beyond conspicuous compassion: Indigenous Australians deserve more than good intentions', Address to the Australia and New Zealand School of Government, Australian National University, Canberra, 7 December 2005, available at: <http://parlinfo.aph.gov.au/parlInfo/download/media/pressrel/GH9I6/upload_binary/gh9i63.pdf>

Weir, JK 2009, *Murray River Country: An Ecological Dialogue with Traditional Owners,* Aboriginal Studies Press, Canberra.

White, N and B Meehan 1988, 'Traditional ecological knowledge: a lens through time', in *Traditional Ecological Knowledge: Wisdom for Sustainable Development,* NM Williams and G Baines (eds), Centre for Resource and Environmental Studies, Australian National University, Canberra: 31–40.

Whitehead, PJ, P Purdon, PM Cooke, J Russell-Smith and S Sutton 2009, 'The West Arnhem Land Fire Abatement (WALFA) project: the institutional environment and its implications', in *Culture, Ecology and Economy of Fire Management in North Australian Savannas Rekindling the Wurrk Tradition,* J Russell-Smith, PJ Whitehead and P Cooke (eds), CSIRO Publishing, Collingwood, Victoria: 287–312.

Yibarbuk, D, PJ Whitehead, J Russell-Smith, D Jackson, C Godjuwa, A Fisher, P Cooke, D Choquenot, DMJS Bowman 2002, 'Fire ecology and Aboriginal land management in central Arnhem Land, northern Australia: a tradition of ecosystem management', *Journal of Biogeography* 28(3): 325–343.

Archival sources

Evans, EC 1963, 'Patrol to establish a road link between Oenpelli and Maningrida', Typescript, Department of Aboriginal Affairs, National Archives of Australia, File 66/3790.

Sweeney, G 1939, Report of patrol in the Junction Bay, Liverpool River and Tomkinson River areas, July-August 1939, National Archives of Australia, File 64/2231.

Legislation

Aboriginal Lands Rights Act Northern Territory 1976 (Cth)

3. The *Kalpurtu* Water Cycle: Bringing Life to the Desert of the South West Kimberley

Patrick Sullivan, Hanson Boxer (Pampila), Warford Bujiman (Pajiman) and Doug Moor (Kordidi)

Water is neither a commodity nor simply an element for the Walmajari people of the Great Sandy Desert fringe. It links the people and their livelihoods directly to the creatures of myth and ritual, the *Kalpurtu*, who have enlivened the natural landscape since the beginning of time. This chapter lays out some of the information on the cultural importance of water which Sullivan recorded during fieldwork organised by Boxer, Bujiman and Moor during the hot dry season of October 2000. The field visits took place over a period of about one week in the vicinity of Yakanarra community, which is about 65 kilometres south-west of Fitzroy Crossing (Toussaint *et al.*, 2001). What follows is a description of the current beliefs and activities of the Walmajari people of this part of the Fitzroy region. It begins with the beliefs and practices of a fairly recent past. In a later section the modern situation is described. Although the people of Yakanarra community do not nowadays hunt, fish and gather bush foods and medicines in quite the same way of their parents and grandparents, several of the older people spent their younger years living in a traditional manner on the periphery of white owned cattle enterprises. Bush food is still an important part of the community's diet and a significant element of cultural identity. Fresh water sources are still important for their food resources and recreation. They may be vital from time to time, since an individual's survival can still depend on finding water when vehicles break down, bog in sand, or when people scout around on foot from the base of a bush camp. Just as the importance of water in this arid area has not diminished, the belief system and practices that surround it remain strong also. Elsewhere in the Kimberley similar beliefs and practices have formed the basis of claims of existing native title (Vachon, 2006). No native title claim has been made in the Yakanarra region, and it is unlikely that recognition of their title could do justice to the richness of the belief system that underlies it, nor do much to protect the spiritual attachment that members of the community have to their local water sources. On the contrary, this chapter shows the wide gap between a system of property based on registered title rights, and a holistic cultural system based in nurturing the natural environment through a combination of appropriate spiritual practices and the use of practical local knowledge.

The Walmajari people of Yakanarra community live on the flood plain of the Fitzroy River and its tributaries. In the wet season, streams flood down off the Saint George Ranges filling the creek systems that feed into the Fitzroy. Swollen with waters from its catchment in the central Kimberley, at the height of the wet season the Fitzroy escapes its banks and floods back onto the plains mingling with the water coming down from the Saint George Ranges. Much of the land goes under water. Movement is restricted and settlement limited to well drained areas of high ground. As the wet season abates the creeks dry out to a series of billabongs, claypans (*pindi*) gradually shrink, the billabongs themselves eventually give out and both people and animals rely on the few permanent rockpools in the ranges, and the soaks (*jila*) in the sandhills of the plains. To the Walmajari people the rain is 'god-given' in more than a simple poetic sense. It is the result of the interaction of the people and the first mythic beings of the Dreamtime – the *kalpurtu*. All of these water sources are linked in Aboriginal culture by the actions of the *kalpurtu*. *Kalpurtu* inhabit the *jila* – permanent spring waters called 'living water' in Aboriginal English or Wunggur Ngaba. The description of cultural attachment to water in this chapter shows how the *kalpurtu* are the centre and source of all water and the animal species associated with it.

Kalpurtu jila, living water

The very first beings were the *kalpurtu*. They are both man and serpent. In this region they were sent out from their centre of origin at Paliyira a Walmajari *jila* west of Yakanarra. One *kalpurtu* in particular defines the region of Yakanarra community and its peoples. This is Moankanambi. In his early journeys he arrived at Pelican Billabong (the original Yakanarra, close to the Old Cherrabun station homestead). Here he looked for a deep water *jila* to settle in. He drew a boomerang out of his stomach, in the manner that *mabarn* or 'witchdoctors' can, and threw it three times in three different directions in search of water. Each time it returned to him since the water it found was not deep enough. In the process it made the flood plains that are cut by Gap Creek (Kungurrmin), Cherrabun Creek (Mankurin) and Christmas Creek. On the fourth throw it sank into the *jila* called Moankanambi near Mona Bore on Go Go station, and the *kalpurtu* knew that this would be his resting place. He journeyed there and that is where he remains. When the hot season builds up and clouds appear in the sky over the desert it is Walmajari practice to dig out the *jila* in a ritual manner to release the water and to make rain.

The *jila* that hold *kalpurtu* are normally dangerous to approach. The *kalpurtu* take the form of snakes with long beards. They are not like any other Dreamtime snake that may also be a known species, for instance a King Brown or a Carpet

Snake. The *kalpurtu* is a unique snake. If the *kalpurtu* is disturbed it can bring misfortune or death. It is important to approach the *kalpurtu* singing the correct song for the particular *kalpurtu* and his *jila*. Walmajari will emphasise that it cannot be just any song, nor can it be a made up song or 'dreamed' song that may be appropriate in other contexts. It is the song given to the people from the Dreamtime, and it makes the *kalpurtu* happy to hear it because he knows he has not been forgotten.

Rainmaking in the hot season

The *kalpurtu* lies deep underground in the waters of the *jila*. In a sense, the *kalpurtu* is the water of the *jila*, just as he is the rain that the people invoke him to produce. In the hot season, lying in his *jila*, he spits out vapour to make the clouds. There are several kinds of clouds. *Mayilbu* is low and near the ground and signals the need to start the rainmaking ritual. *Nangkali* clouds hold *kalpurtu*. The *Kudukudu* clouds contain the seeds of food species that will wash into the ground with the rain and grow and multiply. When the clouds lie close to the earth it is the right time to make rain. The men gather rain stones (*punu*) that are to be found in the vicinity of the *jila*. These stones are distinguished from the others on the ground by the scrutiny of the expert old men. They are taken to the *jila*, where a ritual is performed that is secret. Then the men dig while singing to the *kalpurtu* and the women sing in a group to one side. When the lightning starts to crack in the sky they switch to the lightning song. This way the lightning knows that they are singing to make rain and will avoid striking them, so they are not afraid. Possibly at the same time that the men sing and dig and cover themselves with the clay of the *jila*, or possibly a few days later, the rain comes. The women beat the air and the ground with bushes and cry out to the *kalpurtu* to give them plenty of food in the coming season.

The *kalpurtu* at Moankanambi has other *jila* and a chain of claypans (*pindi*) as part of his complex of water sources. The *jila* called Lumarta, which is some distance away, belongs to him and is linked in local Walmajari conception to the main site. The *pindi* have their own song which is performed while digging out the *jila* and also when visiting the *pindi*. The song asks the *kalpurtu* to fill the *pindi* with good food such as frogs and goanna.

Kalpurtu rain brings food

Walmajari people will explain that the rain contains invisible seeds that go into the ground. These produce, in a few days or weeks, the animal species associated with water – goanna, frogs, land crabs (*kalbagor*), fresh water eels, turtle, fish, ducks, as well as non-food birds such as *pidrureri* (not identified) and the *panjur* which is a seagull that arrives in the desert at this season. With the crabs, eels and turtles this bird gives a strange and dissonant echo of the sea in a land usually seen as the opposite. This echo can be seen also in the presence of the remains of large sea shells, conch and baler, found on the ground surrounding *jila*. These are said to have been left by *kalpurtu*, but can also be seen to be the result of trade routes from the coast into the desert. This, too, probably accounts for the common finding of ground stone axe heads of black granite in the vicinity of *jila*. These do not originate in the region.

The deep, cool underground water which is animated by *kalpurtu* thus gives birth to the cycle of water, beginning with the clouds, then the rain that bears the seeds of food species, and necessarily the running creeks, the billabongs, claypans and rockpools. These various water sources all have their particular qualities and uses, much of them dependent on season.

Ways of catching fish and water borne food species

At the very height of the wet season mobility is affected and water sources cannot be easily used. Yakanarra people are confined to areas of high ground as tracks are unusable. This was one of their reasons for moving from the first outstation at Old Cherrabun station. The creek rises in the wet and floods the plain making the old settlement an island on which the community would be marooned for some time. As the water subsides, the creeks and waterholes become accessible once more. Nowadays the people go fishing with nets and lines. The common practice is to get bait with throw nets and then to set hand lines. When approaching a fishing spot men and women will call out to *kalpurtu* and sing his song asking for a plentiful catch, since ultimately it is the *kalpurtu* who has put the food in the water sources. Fish is an important part of the diet, particularly so during mourning periods when meat is forbidden.

In the past fishing was often a group exercise. In the full flood of the wet season, trees and broken branches are swept along the rivers binding together then snagging in the creek bed and serving as a trap for water-borne sand. As the floods abate, islands and sand banks appear, creating channels at first which gradually dry out to discrete pools and billabongs. Using these channels, sand banks and the banks of the river, groups of men and women would work together pushing bundles of tied branches, leaves and grasses through the water, herding and trapping whatever could be found. According to older community members who participated in these drives, an abundant variety of food would be heaped onto the banks in this way, including small crocodile, barramundi (*pulka*), bream (*tjampinpurra*), catfish (*kulumatjardi*), and freshwater sawfish (*Pristis microdon*).

Another method more suitable to discrete pools is to poison the fish with the bark of the *majarla* tree (*Barringtonia acutangula*, freshwater mangrove). This tree grows conveniently along the banks of rivers and is very commonly found. In the days of stone axes the trees would be cut down and carried on the shoulders of several men to an appropriate pool. Nowadays it is possible to cut the tree into suitable logs *in situ* with a steel axe and transport these to the pool. These poison logs are called *limara*. Normally, cutting the logs occurs the day before their use, they are left overnight and used the next morning. The correct name for the *limara* was given by *kalpurtu* and in their cut state they are associated with the Warlungarri Dreaming. This starts in the north on Oobagooma station then travels to Yeeda and along the Fitzroy to Noonkanbah, then to the east where it terminates in Sturt Creek. The *kalpurtu* who gave the people the technique of using *limara* was a *mabarn* or 'witchdoctor' who also taught the Warlungarri dance. This is a public dance of men, women and children that is widespread in the Kimberley. It precedes initiation ceremonies and is sometimes performed at public festivals simply for amusement.

The fish poisoning technique is most effective when there is a good sized group of people who sit around the edges of the pool, half submerged, and pound the bark of the logs in the pool, eventually stripping it from the log and throwing it into the water until the pool is evenly covered. The bark produces a soapy substance, the frothy water that results is called *jangarla*. The degree of saturation of the pool and the even coverage of the poison is judged by the red stain that spreads from the bark. When this is sufficient the fishers retire to wait. After an hour or two the fish rise to the surface, first jumping to escape the pool, then expiring and floating on the surface where they are easily gathered.

Barramundi and bream can be caught in this way, but during this visit only the bony *lagarr* fish surfaced. While the flesh is good, the hundreds of tiny hair-like bones make eating them a trying experience. The technique that Walmajari people developed to deal with this problem and make use of even this food source is to cook the fish in the normal way over coals, then to let it dry in the sun for one or two days. It can then be pounded into powder with a rock, mixed into a paste with water and eaten.

Seasonal use of water sources

Some pools in the major rivers such as the Fitzroy remain throughout the wet season. These can have mythological significance. One such is Parrakapan on the Fitzroy River near Old Cherrabun homestead. This was a favoured site for spearing crocodile in the station days. This pool has a *kalpurtu* as does another related waterhole on the Forrest River in the region of Jubilee Downs. These *kalpurtu* were put in the river pools by the Dreamtime figure Wurnyambul. He was himself a *kalpurtu* but was also a bird, a black bird which makes a sound like a man. He came from over the Leopold Ranges. He killed two snakes with a stick and when he called their names they made the two rivers. They persist in the permanent water of their respective waterholes. These are said to be 'living waters' like a *jila* but in the creek. Other waterholes that dry out are called *jumu*. This story, like the lore associated with *limara* logs shows how *kalpurtu* belief shades into more common Dreamtime or *ngarangani* stories and overlaps with the northern belief complex centred on *wungurr*, again a mythological snake standing for the creative force itself, but with different characteristics to those of *kalpurtu*.

Billabongs – *jumu*

There are few such large permanent waterholes. Most billabongs will dry out. Kurrkarra billabong in Gap Creek (Kungurrmin) is a good example of this. This billabong was a favoured meeting place between the bush people and the station hands. Here they would become acquainted with flour, tea and sugar, station rations shared by their more acculturated kinsmen. This rapidly replaced *nargati*, the flour made from grass seed. Several older Walmajari, among them Pajiman, one of the authors of this chapter, grew up here before their parents decided to venture life at Old Cherrabun station. Driftwood in the dried up billabong attests to the power of the water in flood. It becomes a useful source of fuel. When the billabong dried up the people would retreat to the rockholes in the ranges – Bulany, Maraltjidi, Tjinan, Karninantjadi, or else they would spread out onto the plains and rely on the water of *jila*.

Claypans – *pindi*

By this time the water in the chain of *pindi*, the claypans that have been filled with *kalpurtu* rain, has also dried out. The *pindi* are said to belong to the *kalpurtu* of Moankanambi as part of his complex of *jila*. When the main *jila* is dug out special songs for the *pindi* are sung. These have been given by *kalpurtu* from the Dreamtime. He is called upon to make the *pindi* deep and rich with food. The songs are again sung when the *pindi* are visited. This is done to make the *kalpurtu* happy, even in the dry season when no water or apparent life remains. Leaving the *pindi*, the people would rely on the rockholes and *jila*.

Rockholes, springs and soaks – *jila*

Bulany is a spring-fed rockhole in the range above Kurrkarra billabong. It is nowadays an important recreation site for the community. Children learn to swim here and the school teachers like to camp. It is popular for fishing. Fresh water eels can be found here. Within walking distance are other rock pools but these are not permanent. However, the presence of water in the ground encourages lush vegetation which produces a micro-climate among the rocks and is a source of food and shade to animals and birds. These springs, like the *jila* in the sand hills of the plains, are favoured hunting locations in the dry season. Attention turns from fish and water based food species to hunting kangaroo and emu. The technique is to hide in bushes at the edge of known watering places, in the old days with a spear, now with a rifle, and wait for the animal to appear. Several *jila* visited during this survey showed tracks or scats of kangaroo and emu. The *jila* become centrally important in the dry season. As Pampila puts it, '*jila* is like a city for our people. It is where everyone goes back to'. This is a fine image because, not only does it suggest Walmajari go 'home' to their *jila* just as white people customarily go home to the cities, it also suggest the comfort, familiarity and the life-supporting aspects of *jila* to the local people.

The word '*jila*' is also given to soaks that are found and dug out in dry creek beds. One such is at the junction of Gap Creek (Kungurrmin) and Cherrabun Creek (Mankurin). Knowledge of the location of these was essential for survival in the past and remains important local knowledge, since exposure to the heat without water for whatever reason poses serious risk. This knowledge has usually been passed down through the generations, but knowledge of the location of soak waters can also be divined in dreams, or more practically by listening to bird song and seeking out where certain birds congregate. Because the *jila*, rockholes and fishing places are used cyclically by the local people as seasons change, they are also sites of shared memory and history. Visiting these places is not just a practical matter of finding water, but also an occasion for remembering birth places, significant life activities and the death of previous generations.

Communities, land, food species and water

In the past, in the long hot period before the first rains, the people would spread out across the landscape in small groups capable of exploiting the existing water resources without completely depleting them. When the season was right they would begin the cycle once more with rainmaking ceremonies celebrating the *kalpurtu*. Nowadays, the bores that have been put down in traditional water sources offer more plentiful and reliable supplies, and make possible the developed communities with their schools and stores that dot the landscape of the Fitzroy River system, even on the desert fringe, such as Yakanarra and Djugerari. These communities have each developed from their own unique history and each has its own cultural circumstances. Yet they are part of the same broad process of development of the Kimberley and share both cultural and practical concerns that arise from this.

Go Go, Cherrabun and Christmas Creek pastoral stations were settled by the Emmanuel brothers in the last decades of the 19th century. The white people and their stock began to compete with Walmajari people for access to water. There was often conflict over spearing of stock that came to waterholes to drink. There was also a belief among the pastoralists that the presence of Aboriginal people at water sources deterred the cattle, which lost condition. The early period of violent conflict gave way to a period of accommodation in which local Aboriginal people worked on the stations during the dry season and went back to the bush for the wet. These station workers encouraged their kin, who remained at large in the desert to the south, to come and settle at the stations. Several older people in the Fitzroy Crossing region remember walking into the stations with their parents after leading a traditional life in the bush. The people who grew up on the stations retained the use of their languages and their knowledge of the land with its sacred and everyday stories.

During the assimilation period in Western Australia many Aboriginal people left the stations for a variety of reasons. The pastoralists began to rely more heavily on paddocking and machinery for stock work and employed smaller work forces. This happened at the same time as they were required to pay their workers cash wages. As the older pastoralists left the industry the new owners of stations tended to run their properties like the farms in the more settled areas, where they exercised exclusive possession. The Emmanuels compromised by allowing the excision of two areas on Go Go and Christmas Creek stations on which the Department of Aboriginal Affairs built housing for dislocated station workers. These are Bayulu and Wankajungka communities, respectively. Many Aboriginal people migrated to the town of Fitzroy Crossing, then found they were prevented from returning to their homelands. The 1970s was a time of great upheaval for most Kimberley Aborigines who found themselves spiritually divorced from their lands as well as materially disadvantaged in town camps and cramped housing developments like Bayulu. The period culminated in a movement for land rights legislation similar to the Northern Territory, but the West Australian legislative proposal failed in Parliament in 1984. In its place the Aboriginal Community Development Program brought a guarantee of Commonwealth funding for outstation infrastructure in tandem with a state government commitment to excise areas of pastoral stations for small Aboriginal homeland communities. This program changed the nature of the Fitzroy Valley. Eight communities have been established on Go Go station alone. These new settlements have developed new relationships with the water sources of the land, while also offering the opportunity for the preservation of traditional cultural values.

The presence of adequate water is a pre-condition for establishment of these small homeland communities. In a sense, Aboriginal people have once more found themselves in competition with pastoralists for water sources, since many excision applications have been rejected or modified because they required existing stock watering places such as bores. The original Yakanarra group first established themselves at Old Cherrabun station, where they had grown up but which is now abandoned. They moved to Mona Bore, Moankanambi, because the Cherrabun area is inundated in the wet season. They were persuaded to move back into Fitzroy Crossing by the Department of Community Development because their presence was thought to be interfering with stock, but then returned to the vicinity once more and set up camp some five or six kilometres from Mona Bore in 1989. Initially they carted water from the running bore to the camp. Later their own bore and a small water tank were installed. More recently a tank of much larger capacity filled by a solar powered pump has been erected. The provision of power, water and housing enables a well serviced community on the people's traditional land. There is also a non-government school. The community now numbers about 150 people. Contemporary relationships to the land and waters are necessarily different after a century of European occupation and modern living conditions for the Aboriginal community. Nevertheless, there is a strong relationship between the cultural practices and beliefs described in the first part of this chapter and the community's contemporary concerns.

The principal effect of abundant water supplies in bores and water holes is the possibility of over stocking the country with cattle, and subsequent environmental degradation. There are signs that this is a problem in this part of the Fitzroy catchment. Stock wander into the creek systems in search of water, destroying river banks which subsequently wash away in the wet season. Over-pasturing leads to bare, dry, dusty plains. Lack of bore maintenance allows the waste of huge quantities of underground water annually. Some of the running bores in the vicinity of Yakanarra have been in operation since the earliest days of the stock enterprise. They have been placed in vicinity of *jila*. It seems very likely that the water that runs off and sinks into the surface does not replenish the underground source. While this is probably very large, it is not a completely inexhaustible resource and such a waste could easily be rectified.

Conclusion

The people of Yakanarra continue to show their commitment to their ancestral lands through living on them and maintaining a highly viable community. It is clear from this description of cultural knowledge of water sources that they could also demonstrate the necessary criteria for recognition of their native title through occupancy and the maintenance of traditions. However, recognition of native title alone would not meet their current needs. The environmental degradation witnessed during this survey is linked in Walmajari thinking to the lack of proper cultural maintenance of water sources. There are fewer and fewer people who know the old rituals with each passing year. This is despite the fact that the young people are growing up understanding their own languages, familiar with the topography of their traditional environment, and at a suitable age participating in initiation ceremonies with other communities. Many cultural practices do survive, even if in a modified form. Nevertheless, there is a danger that some of the more local traditions and stories could disappear. Certainly, the relative scarcity of bush foods and fish, compared to the mature people's memories of their youth, is attributed to recent neglect of the *kalpurtu*. One author of this chapter (Boxer), along with other Walmajari people, has a vision of once more digging out the *jila* of Moankanambi, performing the correct secret ritual, singing the songs that make the *kalpurtu* happy, bringing rain and with it replenishing the earth. This would require bringing knowledgeable people from other communities, and would be an important experience for the youth of the community.

To the Walmajari way of thinking recognition of native title is without content if these activities cannot be performed. Yet recognition would not of itself provide the necessary resources and organisation to revitalise the rituals that Walmajari people believe are necessary to regenerate the land. One way of dealing with both Walmajari and mainstream priorities would be to develop a program of two-way environmental knowledge-sharing. It would be an opportunity to more clearly understand Aboriginal thinking on the nature of water and the regeneration of the land, and therefore the values that they would wish to see preserved in any proposals for water use. Secondly, it could lead to a program of conservation of water sources and monitoring of the use of the land that could produce the regeneration that the people so much desire. This program would involve members of Aboriginal communities themselves in active cross-cultural pursuits that would produce useful results for both Aboriginal and non-Aboriginal users of the land. A necessary first step would be an audit of pastoral station water sources undertaken by community members with local knowledge. Conservation measures for rivers and other water sources would be identified. At the same time appropriate cultural activities and practices could be renewed with particular involvement of young adults. This would result in useful exchange of information between Aboriginal and non-Aboriginal culture with the potential for important practical outcomes. Commonwealth government initiatives such as declaration of an Indigenous Protected Area coupled with employment under the Working on Country could help preserve both the natural ecology and cultural knowledge. On its own, the recognition of native title rights cannot produce these outcomes. They require an alignment of priorities between settler and Aboriginal stakeholders, and a commensurate commitment of resources. This chapter will have achieved its purpose if it serves to lay the ground work for this important area of cross-cultural ecological management in the future.

References

Toussaint, S, P Sullivan, S Yu and M Mularty Jr 2001, 'Fitzroy Valley Aboriginal cultural values study on water (a preliminary assessment)', unpublished report for the Water and Rivers Commission and the Centre for Anthropological Research, the University of Western Australia, Perth.

Vachon, D 2006, 'The Serpent, the Word and the Lie of the Land: the Discipline of Living in the Great Sandy Desert of Australia', unpublished PhD thesis, Graduate Department of Anthropology, University of Toronto, Toronto.

4. 'Two Ways': Bringing Indigenous and Non-Indigenous Knowledges Together

Samantha Muller

For millennia, Indigenous knowledge practices have been fundamental to sustaining Indigenous livelihoods and remain important in many parts of the world. Increasingly, non-Indigenous scientists are engaging with Indigenous peoples to consider collaborative approaches to natural resource management. Often there are challenges of what Christie (2007) terms legibility, in which some elements of Indigenous knowledge are easily understood, documented and objectified by non-Indigenous scientists with other, less tangible aspects being ignored and marginalised. Berkes (1999: 12) agrees that non-Indigenous scientists 'end to dismiss understandings that do not fit their own; this includes understandings of other [Indigenous] scientists using different paradigms'. State-funded and bureaucratically organised knowledge practices make codified, generalised, quantifiable and transferable knowledges legible, but limit recognition for those components of Indigenous knowledge that are 'singular, non-transferable, tacit and unable to be expressed in words' (Christie, 2007: 86). Much research has focused on the contents of Indigenous knowledge systems, the practical and empirical, and how it can be broken down into 'bite-sized chunks of information that can be slotted into Western paradigms' (Ellen and Harris, 2000: 15), at the expense of a deeper understanding of the epistemology of Indigenous knowledges (Briggs, 2005). Verran (2002) refers to non-Indigenous scientists as 'information hunters' seeking to collect the 'facts' on the respectful assumption that Aboriginal communities are where they can find a reservoir of locally specific knowledge. She cites the Northern Territory Parks and Wildlife Commission (Australia) mission statement that includes 'collecting of Aboriginal knowledge' as evidence of the determination to seek 'facts'. Furthermore, there is often then an assumption that these 'factual' knowledges must somehow be scientifically testable to be accepted (Briggs, 2005). However, Indigenous knowledge is not simply a collection of facts, but a way of life.

Christie (2007: 87) refers to Indigenous knowledges as what makes possible the 'routine practices of everyday life'. He identifies the characteristics of Indigenous knowledge as performative, something you do rather than have; context specific, differing from place to place; owned, protected and accountable as it is governed by laws; collective; responsive; active and constantly renewed and reconfigured. Language is also integral to these practices and gives meaning to all things. Christie (2007) also explains that knowledge is not limited to human agency, with the land and other species revealing and keeping knowledge alive.

Therefore, in contrast to the collection of 'facts' which so often characterises non-Indigenous engagement with Indigenous knowledge, in many contexts explanations are no more important than actions. 'They can teach it, they can tell stories about it, they can sing and dance it but they may have no impulse to explain it' (Christie, 2007: 88). Indigenous knowledge tends to be driven by the pragmatic, utilitarian and everyday demands of life and elements of knowledge, including non-Indigenous sciences, can be incorporated into a hybrid, mediated and continually reworked form (Briggs, 2005). Therefore the notion of a 'pristine' knowledge that is 'untouched' is unrealistic and romantic. There are fundamental differences in the ways that Indigenous and non-Indigenous knowledges are socially constructed (Christie, 1990; Sarewitz, 2004; Briggs, 2005).

An influential tradition in western knowledge is a focus on *separation* as the basis for understanding environments. For instance, the separation of humans from nature, and the separation of nature from culture (Weir, 2008). For many Indigenous systems, knowledge is constructed through understanding connections of species to each other, to people, ancestors, stories, dances, art, science, politics, economics, power, society and the cosmos. It is the *connection* that is used to develop context specific information (Christie, 1990; Briggs, 2005; Christie, 2007). Fundamental differences in constructing knowledge, such as through connection or separation, have significant implications with respect to the construction of power relationships, and the marginalisation of Indigenous knowledges for decision-making in natural resource management. Agrawal (1995) states that the link between power and knowledge needs to be explicit for genuine recognition of the contribution of Indigenous knowledges.

Often, the politics of natural resource management can privilege non-Indigenous science as 'objective' and therefore an instrument of power in the hands of 'experts' (Swift, 1996; Novellino, 2003; Briggs, 2005). Natcher and others (2005) argue that power in research and decision-making for resource management is often controlled through the provider of financial, institutional, and political resources. In their research into co-management institutions in Canada, they found power more often involves the 'determination of whose knowledge is of most value to the management process and how such knowledge is or is not used in decision-making' (Natcher *et al.*, 2005: 246). In their research non-First Nation managers tended to define expectations and norms for management and thus produced a discourse of 'truth' that subjugated First Nation knowledges. Similarly, Palmer (2004) documents the marginalisation of traditional owner knowledge and values in Kakadu National Park, Australia. The Park operates on a co-management basis and traditional owners were seeking alternative fisheries management in the park. Their views were misconstrued and portrayed as 'irrational' by government and fishing lobby groups. Palmer argues that 'this unequal power relationship created through an alliance between science and the State leaves the situated knowledge of [Aboriginal people] with a limited field of authority in the non-Aboriginal domain' (Palmer 2004: 61). These examples demonstrate the inadequacies in which Indigenous knowledges are recognised in broader political contexts.

Increasingly, Indigenous and non-Indigenous scientists are working together to create meaningful and more equitable collaborations between knowledge systems. Studies such as those with the Mutijulu Community in central Australia (Baker and Mutitjulu Community, 1992; Reid *et al.*, 1992; Nesbitt *et al.*, 2001) document the benefits of integrating traditional knowledge with non-Indigenous science for ecological research. Terms such as 'two toolboxes' are used by mainstream institutions in Australia in reference to bringing together non-Indigenous and Indigenous sciences (Jackson *et al.*, 1995). Indeed it is a term evoked by many Indigenous resource managers to invite closer collaboration from scientists and their institutions. The promise of 'two toolbox' approaches has often been assumed to work, with little interrogation of the term in a natural resource management context. This paper focuses on the term 'two ways management', a framework developed by Yolngu people in north-east Arnhem Land, Australia. The concept of two ways management seeks to redress the dominance of non-Indigenous science in natural resource management. This paper considers the ontological challenges of integrating Indigenous and non-Indigenous knowledges and considers the practical and resource allocation implications of this divide. It then considers how institutions need to transform in order to honour both knowledge systems.

'Two ways': tracing the concept

Yolngu people are traditional owners of northeast Arnhem Land, Aboriginal owned land in the Northern Territory, Australia. Yolngu have long established metaphors that provide insight and meaning to life. The *Ganma* metaphor identifies how to mix knowledges equitably, how to achieve meaningful two way collaborations. *Ganma* has many meanings, one of which is a place where fresh and salt water meet and mix. The fresh water and the salt water refer to parallel systems of knowledge:

> Strictly speaking, it relates to the separateness of fresh water and salt water knowledge even at the point where they meet and mix. It is like what some [non-Indigenous people] call a "dialectical" relationship, in which two opposed patterns of ideas complement, interact and relate to one another, but never lose their distinctiveness as separate and opposed parts of one whole. (Yunupingu and Watson, 1986: 6–7)

The meeting of the two waters and currents creates a foam on the surface of the water representing the interaction of knowledges (YCEC, 1995). The theory maintains that the combined forces of the streams strengthen each other and lead to a deeper understanding and truth, balancing between complementary opposites, calling for respect and understanding of each others' ways of knowing and doing (Hughes, 2000). *Ganma* has been used as a mechanism to ensure that Yolngu have an equal and active part in the thinking, planning and management of their community institutions. It is not just to have Yolngu people 'involved' in projects, but that the 'the project cannot proceed without the active participation of Yolngu ... in setting directions' (Yunupingu and Watson, 1986: 7). It aims to develop equal dialogue between Yolngu and non-Yolngu through progression along two *raki* (lines of conceptual development).

The Yirrkala Community Education Centre, worked to develop a *Ganma* Curriculum in the mid 1980s to deliver 'two ways' learning for their Yolngu students. Devlin (2004: 26) identifies the key concepts underpinning the two way learning philosophy as sharing power and acknowledging 'competing knowledge systems. ... The main imperative driving [the two way learning] approach is the concept of equality and mutual respect'. The Yolngu concept 'two ways' contests historical institutional power relationships and attempts to build on commonalities and mutual respect rather than difference (YCEC, 1995). Two way learning has been used synonymously in the Northern Territory (NT) to describe bilingual education since the 1970s. Yolngu support for two way learning philosophies extends beyond bilingual education to many community development management initiatives. It has formed the philosophical foundation for Yolngu land and sea management programs such as Dhimurru Aboriginal Corporation (Dhimurru), a Yolngu institution working to develop two-way natural and cultural resource management with non-Yolngu partners. The challenge for this paper is how to translate the institutional recognition and transformations achieved in bilingual education into an environmental governance context.

Dhimurru as a two way institution

Dhimurru is a community-based natural and cultural resource management agency established by Yolngu traditional owners in 1992. Dhimurru is a Yolngu word for the east wind that brings rain and life to the plants bringing new energy and growth (Dhimurru, 2007a). Dhimurru was created in response to traditional owners' concerns of the environmental impacts of a large non-Yolngu mining population in areas around the mining town of Nhulunbuy. The elders sought to protect their lands and developed a permit system to regulate and manage visitor access. Dhimurru has been successful in developing recognition and respect for Yolngu land and sea management. Dhimurru has jurisdiction over approximately 8,500 square kilometres of land and has formalised its management through declaration of an Indigenous Protected Area (Figure 4.1). Dhimurru manages according to its vision statement by Yolngu elders who state that 'the only people who make decisions about the land are those who own the law, the people who own the creation stories, the people whose lives are governed by Yolngu law and belief' (Dhimurru, 2007b). Dhimurru has expanded its management capacity and programs to include a range of land and sea activities, including development of strong partnerships with government and other organisations (Bauman and Smyth, 2007; Langton *et al.*, 2005; Muller, 2008a, 2008b). These developments have at all times sought to align with Dhimurru's vision statement.

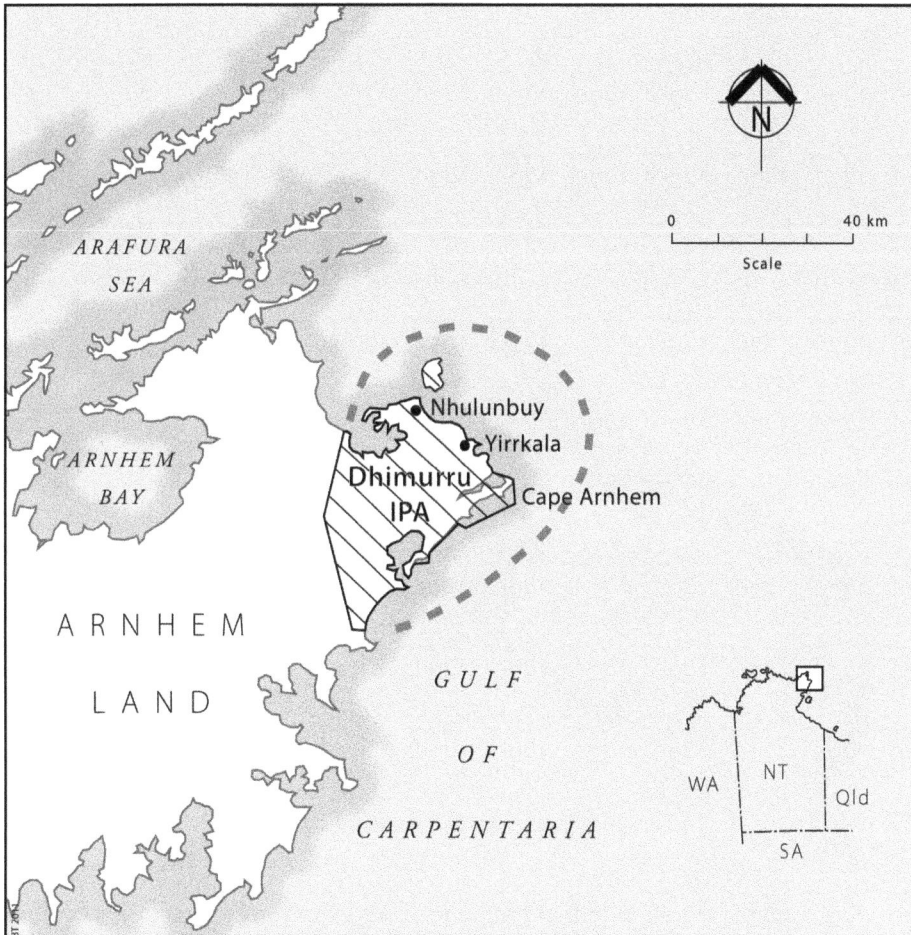

Figure 4.1: Location map of Dhimurru Indigenous Protected Area.

Source: From Muller (2008a).

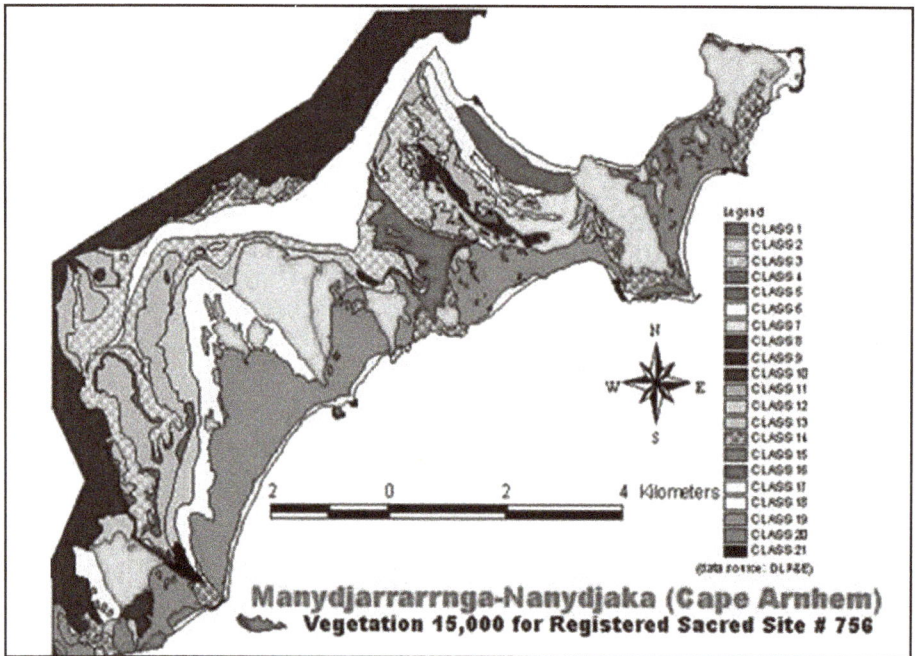

Figure 4.2: Two ways of looking at Cape Arnhem. The picture on the top is a Yolgnu bark painting of Cape Arnhem, showing the connections held between ancestors, plants, animals and people. Below is a cartographic map of Cape Arnhem, which has classified the vegetation into twenty-one different types.

Source: Dhimurru.

The two ways concept has been fundamental in Dhimurru's development of partnerships and programs. At a variety of national and international forums, Dhimurru's Director has used the diagrams in Figure 4.2 to explain different ways of seeing and understanding the same area. Both of the images in Figure 4.2 represent Cape Arnhem, the image on the top is a Yolngu perception of the area:

> When Yolngu look at it, they do not see a painting of animals, they see a map of Cape Arnhem. This is one way of looking at country and how country came to be from a Yolngu side. It shows what wangarr (ancestors) were there and their associations, how people are connected through the land and through animals too. The Western way of looking at Cape Arnhem is different. (Yunupingu and Muller, 2009)

The non-Yolngu perspective draws on western science and categorises Cape Arnhem into management areas as demonstrated in the vegetation categories on the right hand image. Both these perspectives are important in developing appropriate and sustainable management for the area. Dhimurru manages the interface of these knowledges in its development and implementation of coastal management. Bauman and Smyth (2007) document the advantages of the 'two ways' partnership between Dhimurru and the Parks and Wildlife Service (PWS) of the Northern Territory through a formal agreement.[1] The terminology of 'two ways' has been used in a range of other programs and projects that work with Dhimurru and a variety of Indigenous ranger groups (Smith, 2007). Yet despite 'two ways' being projected as an obvious management approach, and its documented successes, there are ontological, logistical, cultural and fiscal challenges which need to be identified and addressed.

Crazy ants as a two way program

The Yellow Crazy Ant Eradication Program hosted by Dhimurru provides an interesting case study to view the application of 'two ways' in resource management. As is the case with the majority of Dhimurru's programs, this project brings together multiple partners under the auspices of 'two ways'. The yellow crazy ant is ranked amongst the world's worst invasive species, having serious environmental, social and economic impacts (Lowe *et al.*, 2000). This ant is known to have inhabited the Gove Peninsula for over 30 years (Majer, 1975) and concern about its presence, coupled with developments in management techniques and products, prompted a project aiming to manage its

---

1 The agreement entails the placement of a Parks and Wildlife Service officer on site at Dhimurru to work in collaboration with Dhimurru rangers. The 21 year agreement provides for a range of technical and managerial support to Dhimurru under the umbrella of Dhimurru's traditional owners governance structure. It was negotiated under section 73 of the *Territory Parks and Wildlife Conservation Act 2000*.

presence in the region. Crazy ants had been identified by external scientists as a significant threat to Yolngu land. Yolngu traditional owners, through Dhimurru as an institution, asserted their involvement as an expression of their authority and obligations under their law to protect and manage their lands. Dhimurru collaborated with the Commonwealth Scientific and Industrial Research Organisation (CSIRO) to co-author an application for funding to manage the ants, with funding and oversight from Alcan mining company, Northern Territory government, Northern Land Council, Indigenous Land Corporation and the federal government through the Department of Environment and Heritage. Structurally, coordination and expertise are provided by CSIRO staff, and most on-ground work is provided by Dhimurru staff. The Crazy Ant eradication program is designed to:

1. establish a community group, the North East Arnhem Land Crazy Ant Management Group (NEALCAMG) to coordinate and implement the crazy ant control program. The NEALCAMG is developing a community based management approach, and sees the capacity building of local land owners as the basis for future pest ant management;

2. increase community awareness about crazy ants and the threat they pose to biological biodiversity, primary industries and society including human health;

3. engage Traditional Owners in north-eastern Arnhem Land to manage and eradicate crazy ant infestations;

4. effectively eradicate crazy ant infestations throughout Arnhem Land, thus eradicating this species from the Australian mainland;

5. protect the significant environmental and cultural values associated with land and country in north-eastern Arnhem Land; and

6. provide national ecological benefits by preventing the escape and spread of this highly invasive species beyond north-eastern Arnhem Land.

Two people are fundamental to the two-way management of this project, the CSIRO non-Yolngu project coordinator, and the Dhimurru Yolngu senior ranger. The project coordinator, Ben Hoffmann, was chosen for his role due to his invasive ant ecology and management expertise, coupled with experience in engaging with Indigenous communities. The senior Yolngu staff member, Balupalu Yunupingu was chosen for his role as he was recognised as a senior cultural authority in the region, coupled with his enthusiasm for the project. This case study aims to document the differences in understandings about the two way nature of this collaborative project between these two key project personnel.

By his own admission, Ben Hoffmann, Crazy Ant Project Coordinator, has struggled conceptually with the implementation of two ways management in the Crazy Ant Project. Ben is focused on the ant itself. As a species new to the region, the ant does not have an associated customary knowledge. Coupled with the apparent minimal role that ants in general seem to have in Yolngu ceremony Ben states that:

> I can't see how Yolngu knowledge is contributing to what is effectively a science-based project … I've struggled to identify how Yolngu knowledge can provide input into the project … I can't find much ant knowledge, and what there is doesn't relate to controlling an invasive species. Because of that, the only 'two-way' thing I can think of, is that the *Ngapaki* [non-Yonlgu] are telling Yolngu what to do. (B Hoffmann, fieldwork interview, Nhulunbuy, NT, 21 October 2005)

The Crazy Ant Project, therefore, does not seem to Ben to provide any opportunity for the inclusion of any traditional knowledge. Like Verran's (2002) 'information hunters' Ben cannot find any 'facts' to include in the scientific management of the project. The invasive species has no place in Yolngu accounts of local environmental relations because it was previously not known in the area. Yolngu participation is essential for ensuring appropriate access and for liaison with the communities which are affected by the project, but 'the project runs according to western science-based and administrative protocols, going out searching for the ant, the way in which we map it and the methodologies for treating it' (B Hoffmann, fieldwork interview, Nhulunbuy, NT, 21 October 2005). Ben does apply himself to learn about Yolngu language, culture and perspectives, however he states that 'immersing myself in the local culture is independent of developing and refining project protocols'. Ben recognises that such cultural awareness and involvement makes it much easier to have the project accepted by the community because he is demonstrating his appreciation of culture, but fails to recognise the Yolngu perspective that this 'cultural education' is an integral part of two-ways management.

For Senior Yolngu Ranger Balupalu Yunupingu, crazy ant work is something new.

> We never done such things like *galkal* [ant] work in our life, you know. This *galkal* is a strange one. We never seen one like this *galkal* before. (B Yunupingu, fieldwork interview, Nhulunbuy, NT, 25 November 2005)

Despite the ant being a new species, Balupalu does not believe that Yolngu knowledge is 'unrelated' to the project. On the contrary, Balupalu feels like he contributes significant knowledge to the project, making it a two way exchange.

> The other way, well I teach him new words from when we doing the survey and all that, I teach him like *wanga* [land] places and all those trees and the *yaku* [names] of the trees. (B Yunupingu, fieldwork interview, Nhulunbuy, NT, 25 November 2005)

For Balupalu, the knowledge of the project is contextualised, encompassing much more than just the ant and the project protocols. The project is also about their personal relationship and so the information they share in the field is important in making it 'two ways'. For example, Balupalu educates Ben in Yolngu language, telling him the names of all the trees and what they are for 'how all the trees are special'. Balupalu even adopted Ben and gave him a Yolngu name to provide Ben a 'place of relatedness' in the Yolngu kinship system. In exchange, Balupalu anticipates being taught Ben's knowledge of western systems as they relate to the project.

Balupalu feels like he has been teaching Ben about the Yolngu system but does not feel that the scale of knowledge-sharing about the systems and structures of the two worlds has been reciprocated.

> No I don't learn about the institutions, that's being keeped [sic] away, from me, you know, and I want to see the whole thing and taught it … I mean I must be recognised as a senior person. I am a senior person in my Yolngu way. I play my part but I like to play that *Ngapaki* [non-Yolngu] way too. (B Yunupingu, fieldwork interview, Nhulunbuy, NT, 25 November 2005)

Ben feels as though he has tried to incorporate Balupalu in the 'higher level' planning and management (B Hoffmann, fieldwork interview, Nhulunbuy, NT, 21 October 2005). Ben referred to a planning meeting he had requested Balupalu to attend at the mine site, 'not only to make him feel involved but also to make sure he really understands how everything's going' (B Hoffmann, fieldwork interview, Nhulunbuy, NT, 21 October 2005). However, just prior to the meeting, Balupalu said he did not want to attend stating 'Ben, what am I going to contribute?' Upon discussion, both agreed that Balupalu's presence at the meeting would predominantly be 'for the sake of being there' and he would feel 'out of place, particularly with all of the mining management' (B Hoffmann, fieldwork interview, Nhulunbuy, NT, 21 October 2005). Balupalu's reluctance to participate in these meetings comes from a feeling that he is not equipped to manage these meetings.

> I'd be just sitting there and asking questions, you know. But in the future, I'd like to be training, so I can get to know and understand language, you know Ngapaki [English] language. (B Yunupingu, fieldwork interview, Nhulunbuy, NT, 25 November 2005)

Balupalu's comments reflect the often discussed issue of 'secret English'[2] (Christie and Perrett, 1996) where Yolngu presume that English languages of power are kept from them. Methodologies employed in the project do not explicitly include learning Yolngu names and culture and language. Neither do they include learning western structures and management frameworks as an explicit goal. However, for Balupalu, the sharing of the big picture worlds is an integral component of the project and 'two ways' management.

The holistic perspective that Balupalu brings to the concept of 'two ways' highlights the expectation that he will learn about non-Yolngu systems and institutions through the project, not just about one species of ant. To Ben, the project is the ant, and its eradication, and therefore he cannot see how Yolngu knowledge can contribute. He believes it is his role to talk to the funding bodies about how they are

> strategically doing it from a scientific point of view. They're not going to listen to a cultural aspect of some project they're investing a million bucks into. (B Hoffmann, fieldwork interview, Nhulunbuy, NT, 21 October 2005)

To Ben, Yolngu language and culture is peripheral to the project. However, Balupalu does not think in terms of a specific ant, but considers the project to also be about the merging of two worlds in which he should have some introduction to the broader institutions of the CSIRO. This analysis highlights the differences in understanding about projects and how they can be perceived to be operating on different scales. The western scientific perspective has decontextualised this project to be all about the ant. However, Balupalu's analysis contextualises and situates the project, sharing all knowledges while they are in the field and recognising that the ant exists on Yolngu land and within a Yolngu context. There are a number of facets limiting cross-cultural, two ways, engagement identified in the crazy ant case study. Underlying these issues is a significant ontological divide between Yolngu and non-Yolngu perspectives. In the following sections I will focus on two issues stemming from that ontological divide that I consider fundamental to limiting effective collaboration. Firstly I will identify the link between non-Indigenous ontological domination and limitation to resourcing Indigenous institutions. Secondly, drawing on bilingual education institutions, I consider the institutional transformation necessary to create ontological equality.

---

2 'Secret English' refers to the claim of many Yolngu that there are secret English languages of power that are being withheld from them. When Yolngu elders make decisions, they employ a range of names and words through their powers of ancestral knowledge and make certain levels of knowledge 'secret' in their production of meaning. Yolngu elders often accuse non-Yolngu people of hiding the 'secret English' of power behind closed doors.

Ontological divides and institutional resourcing

In his discussion of feral animal management in Central Australia, Rose (1995) articulates that:

> Aboriginal people tend to see feral animals as belonging to the country, even though they are recent arrivals. Killing some animals to look after others involves value judgements which are not necessarily part of the Aboriginal world view. (Rose, 1995: 123)

This example highlights the challenges that arise in bringing together alternative ontological perspectives in resource management. There is an assumption that the generation of non-Indigenous scientific research will provide the information needed to respond to ecological and other management issues in 'rational' ways. In the western scientific paradigm it is assumed that everything can be made sense of and systems of categorisation can encompass everything. Therefore western science does not feel it needs to come to terms with Indigenous science as it is considered subsumable within scientific systems. But Indigenous science cannot be reducible to an equivalent in western science; it cannot be translated for it stems from an entirely different ontology (see Weir and Muller, forthcoming). Despite the inadequacy of the non-Indigenous scientific paradigm to address Indigenous concerns, frequently it is presented as the logical and only appropriate response in natural resource management.

> Where agencies are looking to, or being asked to assist in addressing land management issues on Aboriginal land, then they need to be taking some responsibility for addressing those issues in a holistic way. There needs to be understanding that there is more to programs than just the simple objectives of controlling mission grass or controlling crazy ants.
>
> ... you need resources, and you need more than just poison [for spraying weeds] and the fuel that might be required in an ordinary context, because you have to ensure that cultural requirements are being met.
>
> ... Often we [Dhimurru] are the party that needs to pull in the other resources that are needed to address that capacity building, to address the landowner negotiation, address the identification and those issues when we are faced with something like, say, crazy ant eradication
>
> ... I think we would be assisted if there was a greater onus, by Government for example, to address these programs so that we would not face all the responsibility for trying to form those partnerships that are needed to enable that holistic approach to present issues. In other

words the onus shouldn't just be on Indigenous people to try and make this work. (S Roeger, Dhimurru Executive Officer, fieldwork interview, Nhulunbuy, NT, 28 July 2005)

Dhimurru is operating in the interface of 'two worlds' – Yolngu and non-Yolngu. It does so by juggling 30 different contracts, employing 20 Yolngu and six non-Yolngu staff, managing relationships with a range of government and other agencies and ensuring that the Yolngu committee makes all final decisions of importance. One of the challenges Dhimurru faces is securing resources to cover all the costs of its operations. As Steve Roeger, Dhimurru's Executive Officer states, 'if it were just about buying the poison [for spraying weeds] and employing someone to do the job, it would not be as resource intensive as it is' (S Roeger, Dhimurru Executive Officer, fieldwork interview, Nhulunbuy, NT, 28 July 2005). In the case of the crazy ant project, Natural Heritage Trust funds cover the physical costs of the operation. Dhimurru is left to apply to organisations like the Indigenous Land Corporation (ILC) for additional funding to cover effective Yolngu engagement. There is a link here between ontological perspectives and resource allocation. Ben's scientific perspective considers the project to be about the eradication of the ant only and subsequently the funding provided covers that specific focus. However, Balupalu understands the project as being context specific and about the merging of institutions. To him, it is not just a case of managing an invasive ant species, but about learning about Yolngu and mainstream institutions, sharing the context of knowledge production. The Yolngu ontological components are not covered in the grant application, nor are they acknowledged in the other 29 contracts Dhimurru manages.

Non-Indigenous scientific practices that are clearly recognisable as natural resource management are funded by the mainstream resource management agencies, yet what is core business for Dhimurru, managing and valuing the intangible components of Indigenous knowledge, is erased in the funding process. Facilitating effective engagement of Yolngu requires additional resources for ensuring involvement of the right people, consultation with those affected by decisions and capacity building as necessary. It necessitates institutional recognition and resourcing of Yolngu ontologies as fundamental to the success of land and sea management. Dhimurru is managing the operation between two worlds without adequate resources for this specific function.

It was very easy to identify Dhimurru as the organisation who should play the lead role in the project as it was a well established organisation, it had a great reputation for being capable to conduct such work, it had responsibility throughout a lot of the area, and there was agreement by all stakeholders that Dhimurru was most appropriate. (B Hoffmann, fieldwork interview, Nhulunbuy, NT, 21 October 2005)

This quote from Ben demonstrates the value of Dhimurru as a liaison body, a cross-cultural institution and its negotiating role. Implicitly it acknowledges how, as an institution, Dhimurru manages the challenges of operating on Aboriginal land and ensuring effective engagement, employment and consultation with the right people. However, this implicit recognition of the costs involved is not remunerated within the funding agreement. There is no line in the budget for this essential and challenging role. The same is so for the other 29 contracts that Dhimurru has applied for, acquits, manages and continually negotiates ongoing partnerships for (Muller, 2008a). As a result, Dhimurru is under-resourced, and understaffed to manage the elaborate and challenging cross-cultural environment that it works within. The domination of non-Indigenous scientific thinking, protected through non-Indigenous grants administration, marginalises Indigenous knowledges from the process. It requires an institutional transformation to respect and honour both ontologies equally.

Transforming institutions – towards ontological equality

Drawing on the *Ganma* metaphor as the basis of 'two way' learning we learn that the salt water and fresh water are constructed as a separate and different. Yet when the salt and fresh waters meet, they create brackish water with what many Yolngu refer to as a 'different taste'. Importantly, both the salt and fresh waters are necessary for this interaction to occur where neither salt nor fresh is more important, overpowering or dominant in their mixing. Yet, when we reflect on this metaphor in the context of environmental governance institutions in Australia the politics of mainstream environmental institutions overpower Yolngu institutions. Drawing on the example of bilingual education, this section considers how to develop context specific resolutions for reconfiguring power in environmental governance.

Bilingual education opens the opportunity for participants, both teachers and students, to participate in two languages and therefore two ontologies and as such engage in individual and institutional transformations. Bilingual institutions have the capacity (despite significant controversy as to whether the capacity is sufficient or adequately resourced) to deliver outcomes that allow participants to participate equitably in two languages and cultures. From a Yolngu vantage point, the transformation in bilingual education is not complete as meetings with government partners are not conducted in *Yolngu Matha* and remain transacted in English – the language of power. However, due to the specificity of context, it would be unreasonable for *Yolngu Matha* speakers to expect government agents to conduct meetings with central Australian communities in *Yolngu Matha* as communities there speak different languages. Therefore, in the context of whole-of-system discussions, English becomes a *lingua franca* – a shared language. The challenge is for this common tongue to be managed not as a language of power, not as a language of dispossession and colonisation, but as a language of partnership, collaboration, mutual respect and engagement. In monolingual education, English is a language of domination, where *Yolngu Matha* is neither visible nor respected. In bilingual education English is used as a *lingua franca*, where education institutions respect and recognise alternative languages commensurate with necessary resources. Even though English is used as a language of commons, *Yolngu Matha* and English are institutionally recognised and therefore resourced. The challenge identified in this paper is how to translate such a transformation into an environmental governance context.

In environmental governance, we find an incapacity for mainstream environmental governance institutions to respond to, respect or resource the contribution of Yolngu knowledges and practices. For example, Indigenous rangers are systemically inadequately resourced for their provision of environmental services (Luckert *et al.*, 2007, Muller, 2008b). Some institutions and government programs in Australia are seeking to engage with these issues, such as the Indigenous Protected Areas Programme (Smyth and Sutherland, 1996; Muller, 2003; Langton *et al.*, 2005; Bauman and Smyth, 2007), but so far these opportunities are rare and limited. Instead of transforming government institutions, it is Indigenous organisations that have to manage, translate, transform and broker the relationships between their communities and mainstream agencies, without the resources or respect that this role requires. In current environmental management governance non-Indigenous science is the language of dominance. At this point in time there is no *lingua franca* in terms of process and relationship that will allow for respect for difference by government institutions, there is no language to begin to frame this discussion as the bilingual education does. Consequently, as Natcher and others state 'because ethnicity and power are related directly to the visibility of knowledge and its holder, the application of indigenous [sic] knowledges to the management process is most often subjugated against the western ontologies' (Natcher *et al.*, 2005: 247).

One of the difficulties of this situation is that scientists like Ben do not see western science as a language of domination. He adheres to the view that conservation science can be made value free and when done rigorously can transcend any particular cultural context. Interviews with him demonstrate that he is keen to embrace and work with Yolngu knowledge and actively seeks to learn *Yolngu Matha* language and cultural protocols, but cannot see how the project is operating in a two way framework. Ben is seeking 'factual' Yolngu knowledge that can be directly applied within the eradication framework, not the connections, relations and vales of Yolngu knowledge. His understanding does not embrace Yolngu knowledge as characterised by Christie (2007) as, performative, context specific, active and constantly renewed and revitalised. For Balupalu, when work is conducted on Yolngu land it is intimately connected to the Yolngu world. To Yolngu this is obvious; this is what Balupalu takes for granted. He does not see just an ant causing a problem, and if he thinks that is all others see then he sees a lack of respect. Ben has been trying to learn language as a means of respecting culture, but does not see that his insistence on the dominance of non-Indigenous science from which 'factual' knowledge is privileged in the project is what is limiting his engagement and what is necessary for true institutional transformation. The issue of invisibility of power to cultures of power is just as urgent as the invisibility of the rights of marginalised groups. Indeed invisibility of power is what leads to 'deep colonisation' (see, Rose, 1999) whereby, because of the invisibility of Yolngu perspectives, unintentional colonisation of programs and processes occurs. The failure of policies, agencies and bureaucrats to offer acknowledgement of alternative ontologies in the way they construct and fund projects, regardless of their rhetoric, is the ontological arrogance that reveals the real exercise of power. What this paper seeks to challenge is that broader environmental governance institutions need to transform to a point where Indigenous and western science are ontologically recognised and therefore institutionally resourced as equals.

The fact that the Crazy Ant application was co-authored with Dhimurru indicates that there is something missing from all parties understanding about how to achieve meaningful collaboration. Presumably, Dhimurru has the opportunity to insert additional costs for engaging in Yolngu business and management as a budget line. But simply providing the resources will not satisfy Balupalu's desire to learn about western institutions, such as the CSIRO. He seeks to understand what the 'secret English' is that limits meaningful collaboration, to demystify non-Indigenous institutions. This paper argues for more than simply additional financial resources. There is a risk that organisations like Dhimurru may succeed in gaining additional funding without the requisite respect for ontological difference. Non-Indigenous science, as a language of domination, sets the terms of engagement and is used for legitimacy with government funding bodies. Consequently, Balupalu remains dissatisfied. It is not simply additional funds that will meet this satisfaction, but ontological valuation of Yolngu knowledge, process and performance.

Perhaps there is no mechanism to incorporate the unquantifiable or intangible elements of Yolngu knowledge and management approaches within rigid objectives and contractual obligations. Indeed, the common use of the term 'objectives' not only denies the subjectivity of western cultural knowledge, but also limits the valuing and incorporation of other 'subjective' knowledge and practices. In this way, the instrumentality of the grant process is a significant limitation to recognising Yolngu knowledge. 'A different focus on the processes of Indigenous knowledge might therefore generate a deeper and more dynamic understanding of change' (Briggs, 2005: 108). Indigenous knowledge is not a static database of facts, but fluid and constantly changing, open to renegotiation and incorporation of new information (Sillitoe, 1998; Briggs, 2005; Christie, 2007). Balupalu's insights demonstrate that the project has value far beyond the specific species of ant it aims to manage. His knowledge is embedded in the everyday practices of life and he learns knowledge from the country and species in it. His knowledge is performative and constantly able to be renegotiated. Effective engagement of Yolngu knowledge requires a focus beyond the 'facts' or 'outcomes' or 'objectives' of the project towards a valuing of relationships and process and the subjective values. 'The negotiations towards shared understandings and strategies among divergent knowledge systems must be a continuous process within an adaptive framework, rather than a question of specifying a fixed set of indicators' (Christie, 2007: 88). There is a need to develop a *lingua franca* such that non-Indigenous science is not dominant, but complementary to Indigenous perspectives.

Conclusions

Despite its heralded benefits, there remain significant ontological limitations to truly embracing 'two ways' in Indigenous land and sea management. Mainstream natural resource management departments and funding bodies interpret the environment in such a way to preclude funding for Indigenous authority and practice. The domination of non-Indigenous ontological perspectives marginalise Indigenous ontologies and thus limit the resourcing of Indigenous land and sea management. Even when collaboration is developed jointly with good intentions from all partners, there is still dissatisfaction with the collaboration because of the ontological breach. In the Crazy Ant example, funds are provided for the 'objectives' defined by non-Indigenous ontological perspectives, thus resulting in domination of non-Indigenous values and objectives. As a language of domination, non-Indigenous science does not value Yolngu ontologies beyond what 'facts' they can find and therefore there are no mechanisms to incorporate them in funding applications as 'objectives'. Consequently, it is the Indigenous institution, in this case Dhimurru, that is left to manage the holistic context in which these project exist, to facilitate Yolngu engagement, training, employment and the subjectivities of Yolngu knowledge and to meet the costs of this process.

This paper highlights the issues of invisibility of power of dominant cultures and the implications of those power relationships in resource and environmental management. It challenges the dominance of non-Indigenous science in natural resource management seeking to inspire a transformation of institutions in resource management relationships based on recognition and respect of difference. In the hunt for 'facts', the performative and essential characteristics of Indigenous knowledges are not seen, and are therefore marginalised by cultures of power. Following on from this, there is a close relationship between ontological recognition and subsequent resource contribution. This paper encourages all participants to reconsider and rethink approaches to meaningful collaboration in which the non-transferable, tacit and unquantified knowledges are recognised and adequately resourced to create a language of equals between Indigenous and non-Indigenous sciences.

> Unless we do engage with those intangible dimensions or the wider social aspect of Indigenous knowledge – which includes ceremony, kinship, ritual, hunting, harvest, all of those things – and until we engage with them, Indigenous knowledge will continue to be subsumed into mainstream agendas. (Jackson *et al.*, 1995)

Acknowledgements

I would like to thank Balupalu Yunupingu, Steve Roeger and Ben Hoffmann for sharing their perspectives. A special thanks to Ben for his ongoing constructive discussions about science and ontological challenges during various drafts. Thank you also to Prof Richie Howitt, Dr Sandie Suchet-Pearson, Greg Wearne and Dr Sue Jackson for their invaluable time, assistance and insight to various drafts of this paper.

References

Agrawal, A 1995, 'Dismantling the divide between indigenous and scientific knowledge', *Development and Change* 26: 413–439.

Baker, LM and Mutitjulu Community 1992, 'Comparing two views of the landscape: Aboriginal traditional ecological knowledge and modern scientific knowledge', *The Rangeland Journal* 14(2): 174–189.

Bauman, T and D Smyth 2007, *Indigenous Partnerships in Protected Area Management in Australia: Three Case Studies*, Australian Institute of Aboriginal and Torres Strait Island Studies, Canberra.

Berkes, F 1999, *Sacred Ecology: Traditional Ecological Knowledge and Resource Management*, Taylor and Francis, Philadelphia.

Briggs, J 2005, 'The use of Indigenous knowledge in development: problems and challenges', *Progress in Development Studies* 5(2): 99–114.

Christie, M 1990, 'Aboriginal science for the ecologically sustainable future', *Ngoonjook: A Journal of Australian Indigenous Studies*, November: 56–68.

— 2007, 'Knowledge management and Natural Resource Management', in *Investing in Indigenous Natural Resource Management*, MK Luckert, B Campbell and JT Gorman (eds), Charles Darwin University Press, Darwin: 86–90.

— and B Perrett 1996, 'Language, knowledge and the search for "Secret English" in northeast Arnhem Land', in *Resources, Nations and Indigenous Peoples. Case Studies from Australia, Melanesia and Southeast Asia*, R Howitt, J Connell and P Hirsch (eds), Oxford University Press, Melbourne: 57–65.

Devlin, B 2004, 'Two-Way Learning in the NT: Some Research Based Recommendations', Report prepared for the NT Department of Employment, Education and Training, Charles Darwin University, Darwin.

Dhimurru 2007a, 'Dhimurru website', available at: <www.dhimurru.com.au> (accessed 22 May 2007).

— 2007b, *A Visitor's Guide: Recreation Areas – North East Arnhem Land*, Dhimurru Land Management Aboriginal Corporation, Darwin.

Ellen, R and H Harris 2000, 'Introduction', in *Indigenous Environmental Knowledge and its Transformations*, R Ellen, P Parkes and A Bicker (eds), Harwood Academic Publishers, Amsterdam: 1–33.

Hughes, I 2000. *Indigenous Knowledge for Reconciliation and Community Action*, Participatory Action Research World Congress, Ballarat.

Jackson, S, DB Rose and S Johnson 1995, 'A burgeoning role for Aboriginal knowledge', *Ecos* 125 (June-July): 10–13.

Langton, M, ZM Rhea and L Palmer 2005, 'Community-oriented protected areas for Indigenous people and local communities', *Journal of Political Ecology* 12: 23–50.

Lowe, S, M Browne, S Boudjelas and M De Poorter 2000, '100 of the world's worst invasive alien species. A selection from the Global Invasive Species Database', *Aliens* 12: 1–12.

Majer, JD 1984, 'Recolonisation by ants in rehabilitated open-cut mines in northern Australia', *Reclamation and Revegetation Research* 2: 279–298.

Luckert, MK, B Campbell and JT Gorman (eds) 2007, *Investing in Indigenous Natural Resource Management*, Charles Darwin University Press, Darwin.

Muller, S 2003, 'Towards decolonisation of Australia's Protected Area management: the Nantawarrina Indigenous Protected Area experience', *Australian Geographical Studies* 41(1): 29–43.

— 2008a. 'Accountability constructions, contestations and implications: insights from working in a Yolngu cross-cultural institution, Australia', *Geography Compass*, available at: <http://www.blackwell-compass.com/subject/ geography/article_view?highlight_query=muller&type=std&slop=0&fuzz y=0.5&last_results=query%3Dmuller%26topics%3D%26content_types% 3DALL%26submit%3DSearch&parent=void&sortby=relevance&offset=0& article_id=geco_articles_bpl087> (accessed 14 March 2008).

— 2008b, 'Indigenous Payment for Environmental Service (PES) opportunities in Northern Territory: negotiating with customs', *Australian Geographer* 39(2): 149–170.

Natcher, D, S Davis and C Hickey 2005, 'Co-management: managing relationships, not resources', *Human Organization* 64(3): 240–250.

Nesbitt, B, L Baker, P Copley, F Young and Anangu Pitjantjatjara Land Management 2001, 'Co-operative cross-cultural biological surveys in resource management: experiences in the Anangu Pitjantjatjara lands', in *Working on Country: Contemporary Indigenous Management of Australia's Lands and Coastal Regions*, in R Baker, J Davies and E Young (eds), Oxford University Press, Melbourne: 187–198.

Novellino, D 2003, 'From seduction to miscommunication: the confession and presentation of local knowledge in "participatory development"', in *Negotiation Local Knowledge: Power and Identity in Development*, J Pottier, A Bicker and P Sillitoe (eds), Pluto Press, London: 273–297.

Palmer, L 2004, 'Fishing lifestyles: "Territorians", Traditional Owners and the management of recreational fishing in Kakadu National Park', *Australian Geographical Studies* 42(1): 60–76.

Reid, J, L Baker, S Morton and Mutitjulu Community 1992, 'Traditional knowledge + ecological survey = better land management', *Search* 23(8): 249–251.

Rose, B 1995, *Land Management Issues: Attitudes and Perceptions Amongst Aboriginal People of Central Australia*, Central Land Council, Alice Springs.

Rose, DB 1999, 'Indigenous ecologies and an ethic of connection', in *Global Ethics and Environment*, N Low (ed), Routledge, London: 175–187.

Sarewitz, D 2004, 'How science makes environmental controversies worse', *Environmental Science & Policy* 7: 385–403.

Sillitoe, P 1998, 'Knowing the land: soil and land resource evaluation and indigenous knowledge', *Soil Use and Management* 14: 188–193.

Smith, R 2007, 'Caring for Country Ranger Programs in the Daly Region: Towards Sustainable Futures', unpublished Hons thesis, Macquarie University, Sydney.

Smyth, D and J Sutherland 1996, *Indigenous Protected Areas, Conservation Partnerships with Indigenous Landholders*, Environment Australia, Canberra.

Swift, J 1996, 'Desertification: narratives, winners and losers', in *The Lie of the Land: Challenging Received Wisdom on the African Environment*, M Leach and R Mearns (eds), International African Institute, London: 73–90.

Verran, H 2002, 'A postcolonial moment in science studies: alternative firing regimes of environmental scientists and Aboriginal landowners', *Social Studies of Science* 32(5–6): 729–762.

Weir, JK 2008, 'Connectivity', *Australian Humanities Review* 45: 153–164.

Weir, JK and SL Muller (forthcoming), 'Caring for country is not natural resource management'.

Yirrkala Community Education Centre (YCEC) 1995, Curriculum development in Yirrkala CEC and Yirrkala Homelands Schools, Yirrkala.

YCEC – *see* Yirrkala Community Education Centre

Yunupingu, B and H Watson 1986, 'The Ganma project in mathematics curriculum. A draft proposal outline', Discussion paper for the Laynhapuy Association, Yirrkala Community School and Homelands Centres Schools.

Yunupingu, D and S Muller 2009, 'Dhimurru's Sea Country Planning journey: opportunities and challenges to meeting Yolngu aspirations for sea country management in Northern Territory, Australia', *Australasian Journal for Environmental Management* 16: 158–167.

5. Water Planning and Native Title: A Karajarri and Government Engagement in the West Kimberley[1]

Jessica K Weir, Roy Stone and Mervyn Mulardy Jnr

We're a coastal tribe, but we live in a desert area. We depend almost exclusively on the rain that comes down and the water that's holding the groundwater table. … If our soaks and springs start to dry up, then the trees will start to die off and the animals will start to die too. Because we do live in the edge of the desert and in the desert area, the water is so important for us for our living area. We'll continue to negotiate with all the other stakeholders in our country, and do the work that's required, so we get the proper benefits out of all of that. (J Edgar, Deputy-Chair Karajarri Traditional Lands Association, comment from the floor, National Native Title Conference 2008, Perth, 5 June 2008)

With prolonged drought and climate change, water has once again come into focus in national, state and territory policy development, prompting the review of water planning and management. As part of this, policy makers have revisited the complexities of establishing large-scale industrial agriculture in Australia's tropical monsoon country (Ross, 2009). Water planning in Karajarri country in the West Kimberley of Western Australia is occurring within this context. For Karajarri, they hope that the planning process will help ensure that water is treated the 'right way' for country (Mulardy Jnr cited in Mathews, 2008).

Karajarri have an important water story to tell. Their life in and on the fringes of the Great Sandy Desert has always depended on the flowing fresh groundwater that is also called the La Grange Basin (Figure 5.1). Karajarri describe this water as 'living water', and relate powerful stories to how the water moves through country (Yu, 1999). In the late 1990s a large scale cotton irrigation proposal threatened to upset these relationships. At the same time, Karajarri were in the process of having their native title rights recognised (Figure 5.2). The irrigation proposal lead to a range of government funded studies into the stakeholder, cultural and ecological values that are supported by La Grange groundwater. The majority of stakeholders held concerns over the impact of such a large

---

1   This paper has developed out of a presentation given by Roy Stone at the National Native Title Conference 2008, Perth, 5 June 2008. The authors thank Joe Edgar, Anna Mardling, Luke Taylor, Tran Tran and two anonymous peer reviewers for their comments on an earlier draft. Our thanks also go to Sarah Yu for her assistance with place names.

irrigation proposal for this small community and arid country. These studies revealed intimate, complex and life-sustaining relationships that thrive on the freshwater, which could be jeopardised by the large-scale extraction of groundwater. Fortunately for the people holding these concerns, the irrigation proposal did not go ahead.

This paper describes the successful local opposition to the cotton proposal, however, this paper is not set around this event. Rather, the event is a meaningful prompt for discussions about how water decisions are being made, how they could be made, and how water planning could be better designed to support development aspirations of the communities within the Basin. In the preparation of a water allocation plan for the La Grange groundwater area. Karajarri and the Western Australian Department of Water have sought to find a common ground through engagement, dialogue and research into water management and understandings of water. This plan has now been finalised as the La Grange groundwater allocation plan (Department of Water, 2010).

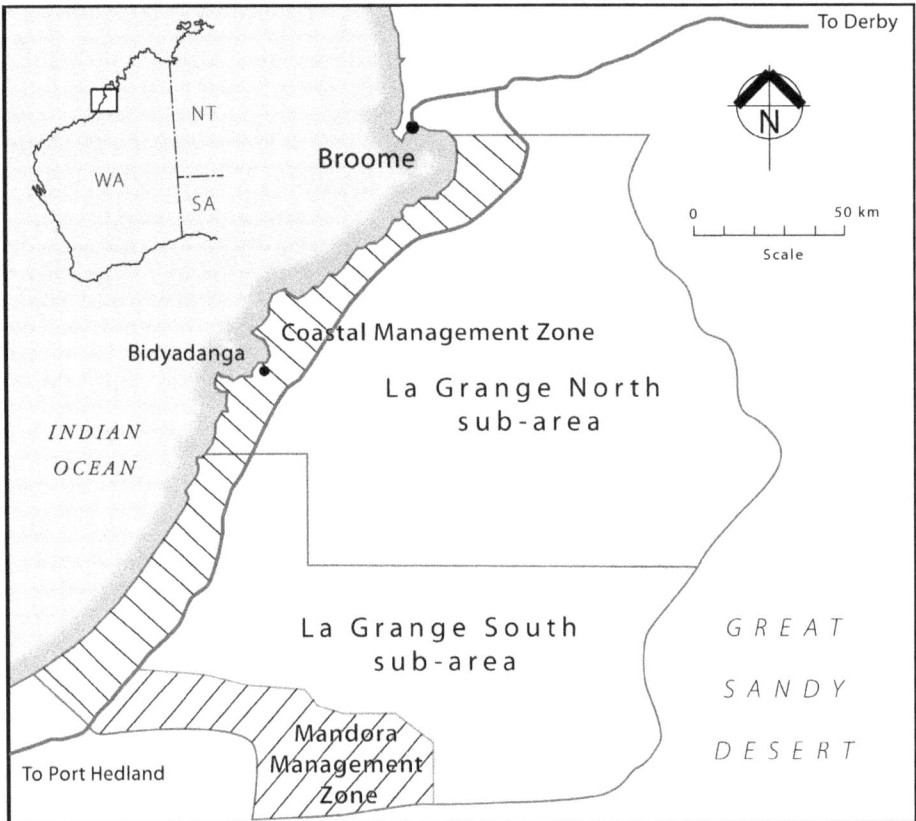

Figure 5.1: Potential management zones for the La Grange sub-basin.

Source: Adapted from Roy Stone, PowerPoint presentation, National Native Title Conference, Perth, 5 June 2008.

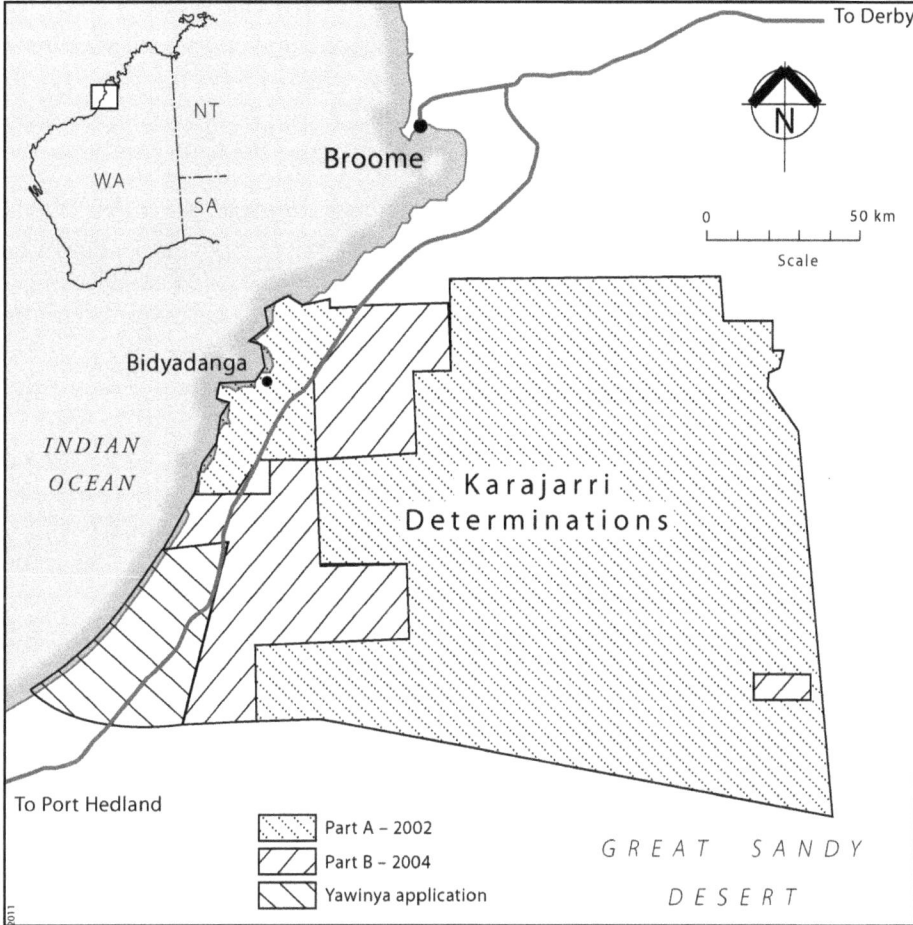

Figure 5.2: Karajarri 2002 and 2004 native title consent determinations, and the Yawinya native title application which is in negotiation with Nyangumarta. Part A is chiefly exclusive native title rights, and Part B is chiefly non-exclusive native title rights. Part B includes the inland Dragon Tree Soak Nature Reserve, which was determined as extinguishing native title.

Source: Adapted from National Native Title Tribunal, 2004, and National Native Title Tribunal, 2011.

Cotton and native title

The La Grange Basin is one of Western Australia's (WA) largest groundwater resources. This is the arid country of the Great Sandy Desert, including where the desert meets with the Indian Ocean along the West Australian coastline south of Broome. Prior to this planning process, little water in the La Grange Basin was allocated for consumptive use by humans, with licensed use of water at 1.8 gigalitres per year and unlicensed (stock, domestic purposes, drinking water) at 3.0 gigalitres (Department of Water, 2008: 17). These uses include the domestic water supply for the community of Bidyadanga and some small scale horticulture. Water is also profoundly important to Karajarri for many other reasons in addition to human consumption. The health of country, including many wetlands, plants and animals, are all dependent on the fresh groundwater.

In the 1990s, the private company Western Agricultural Industries (WAI) became interested in the potential of this freshwater to support irrigation industry. Whilst a range of crops of interest were identified (such as sugar cane, leucaena, exotic hardwoods, hemp, viticulture, freshwater aquaculture), cotton production was the main focus. In the late 1990s the WA government called for expressions of interest in the development of large scale irrigated agriculture in the west Kimberley. In response WAI developed the West Kimberley Irrigation Proposal, a two stage development. The first stage relied on an allocation of more than 90 per cent of the La Grange groundwater potentially available for consumptive use to support an envisaged 20,000 hectares of irrigated cotton fields. The second stage would expand this industry by diverting water from the Fitzroy River, which is far to the east of Karajarri country. Fitzroy water would be diverted west and then south by an irrigation canal to support a much larger 200,000 hectares of cotton production. This extraction and diversion of water was designed to take advantage of the free draining soils of the Karajarri coast, which is Pindan country with acacia thickets and low trees. This coastal strip is also where pastoralists have leased land for pastoral stations.

In 1998, WAI signed a Memorandum of Understanding with the WA government to carry out a feasibility investigation of the WAI proposal. Whilst not a legal contract, the Memorandum of Understanding indicated the WA government's intention to support the irrigation industry, dependent on the proposal's feasibility. Because of the size of the WAI proposal, it also triggered initiation of a sub-regional water allocation plan for the La Grange groundwater. The policy of the WA department responsible for water, the Water and Rivers Commission (known today as the Department of Water), held that where a single large development proposal generates the need for such an allocation plan, the development proponent must undertake and fund much of the groundwater investigation. This ensures the government does not meet the feasibility study costs of a private developer (Beckwith and Associates, 1999: 16). WAI was instructed to undertake a feasibility study into ground water use from the La Grange resource and a pre-feasibility study of surface water use from the Fitzroy River. A groundwater investigation licence was granted to WAI for exploration of the La Grange groundwater resource. In 1999, WAI began cotton trials at Shamrock Gardens pastoral station on the Karajarri coast (Beckwith and Associates, 1999: 3).

Whilst the WAI proposal was being investigated, Karajarri were progressing work towards the formal recognition of their native title rights and interests. Their 1996 native title application was initially in response to a new pastoral fence on Shamrock station. The fence extended from a turkey nest dam that had been recently constructed by WAI. Karajarri elder Wittidong Mulardy was concerned because it interfered with Karajarri access to the culturally significant Parturr hills. In addition to their native title application, Karajarri investigated ways to purchase the pastoral leases for the two stations Shamrock and Nita Downs. However, the leases to Shamrock and Nita Downs were purchased by WAI as part of their cotton proposal. Karajarri decided to expand their native title to almost a 'whole of country' application to take a more comprehensive approach to such matters (Yu, 1998: 1; S Yu, pers comm, 7 May 2008). With additional applications made in 1997 and 1999, Karajarri lodged a combined native title application in 2000.

As the native title work progressed, the cotton proposal became intimately interlinked with it. WAI, as owner of the Shamrock and Nita Downs pastoral stations, would need to agree to any native title consent determination covering that area. But WAI in turn needed access to other lands on Crown radical title (unallocated Crown land)[2] to carry out groundwater drilling investigations and this was constrained by the Karajarri native title application. In 2001, WAI decided to stop the La Grange groundwater investigation work until the native title issues were determined.

As part of their native title application, Karajarri prepared extensive evidence for the Federal Court with professional expertise coordinated and provided by the Kimberley Land Council. Their application was opposed in the Federal Court by the State of Western Australia, the Commonwealth government, the Shire of Broome, Telstra, pastoral and agricultural interests, pearling companies and fishing interests, exploration companies and the Bidyadanga Community Council. Federal Court Justice North found that the strength of the applicants' evidence and expert evidence was such that native title mediation would be better than litigation. That is, Karajarri native title should be determined by agreement between the parties.

In 2002 and 2004 two consent determinations were agreed upon by all parties and signed off by the Federal Court. The 2002 consent determination focused largely on Crown radical title, and recognised that Karajarri held exclusive native title rights and interests to 'possess, occupy, use and enjoy' their country 'to the exclusion of all others' (*Nangkiriny v Western Australia*, 2002; see also,

---

2   Prior to the *Mabo* decision (*Mabo v Queensland [No 2]* (1992) 175 CLR 1), Crown radical title was called unallocated Crown land within the Australian property law system. Such lands continue to be commonly described as unallocated Crown land even though the High Court has determined otherwise.

Weir, 2011 and Edgar, forthcoming). In 2004, the parties reached their second consent determination, the majority of which recognised non-exclusive native title rights and interests over the pastoral leases (*Nangkiriny v Western Australia*, 2004). The outcomes of these two consent determinations are summarised in Table 5.1.

Table 5.1: Karajarri native title rights and interests

In the 2002 consent determination, Karajarri are recognised as holding *exclusive* native title rights and interests to 'possess, occupy, use and enjoy' their country 'to the exclusion of all others'. The Federal Court described this as including:

i. the right to live on the land;

ii. the right to make decisions about the use and enjoyment of the land and waters;

iii. the right to hunt, gather and fish on the land and waters in accordance with their traditional laws and customs for personal, domestic, social, cultural, religious, spiritual, ceremonial and communal needs;

iv. the right to take and use the waters and other resources accessed in accordance with their traditional laws and customs for personal, domestic, social, cultural, religious, spiritual, ceremonial and communal needs;

v. the right to maintain and protect important places and areas of significance to the Karajarri people under their traditional laws and customs on the land and waters;

and

vi. the right to control access to, and activities conducted by others on, the land and waters, including the right to give permission to others to enter and conduct activities on the land and waters on such conditions as the Karajarri people see fit.

Most of the 2004 consent determination concerned the pastoral stations Nita Downs Station, Shamrock Station, and part of Anna Plains Station, as well as the De Grey Stock Route and a number of other small areas of land. Karajarri non-exclusive native title interests are recognised as existing in this area. These are:

i. the right to enter and remain on the land and waters;

ii. the right to camp and erect temporary shelters;

iii. the right to take fauna and flora from the land and waters;

iv. the right to take other natural resources of the land such as ochre, stones, soils, wood and resin;

v. the right to take the waters including flowing and subterranean waters;

vi. the right to engage in ritual and ceremony;

and

vii. the right to care for, maintain and protect from physical harm, particular sites and areas of significance to the Karajarri people.

With respect to native title rights to water, the *Native Title Act 1993* (Cth) makes explicit reference to the fact that native title rights and interests extend to both 'land and waters'. Waters are defined as including:

(a) sea, a river, a lake, a tidal inlet, a bay, an estuary, a harbour or subterranean waters; or

(b) the bed or subsoil under, or airspace over, any waters (including waters mentioned in paragraph (a)); or

(c) the shore, or subsoil under or airspace over the shore, between high water and low water. (*Native Title Act 1993* (Cth): s 253)[3]

However, where water is a resource for consumptive use by humans, the *Native Title Act 1993* (Cth) confirms any Commonwealth, State and Territory legislation that asserts ownership of natural resources, and/or the right to use, control and regulate water (*Native Title Act 1993* (Cth): s 212). The High Court has argued that this vesting is sufficient to prevent exclusive native title possession of the water, with only the right to take remaining (Strelein and Weir, 2008; see also, for the High Court's findings, *Western Australia v Ward*, 2002: para 263; and, for an apparent contrast to these findings, see the High Court decision in *Yanner v Eaton*, 1999). That is, the native title holders only have usufructuary or use rights, not ownership rights, to the water. Thus, the legal framework for native title and water emphasises the continuation of the current water governance arrangements with limited opportunity for native title holders to

3   Land is defined as 'the airspace over, or subsoil under, land' and does not include 'waters', *Native Title Act 1993* (Cth): s 253.

make decisions about water allocations that could have a major impact on their rights and interests. Yet without water, Karajarri would not be able to enjoy or exercise the majority of their native title rights and interests.

In addition to this legal limitation, Karajarri native title rights and interests in water are further constrained by the details of their consent determinations. Every native title determination is different, reflecting the particular traditional laws and customs of the native title holders, and the types of land tenures and legislation in the determination area. Where native title is reached through consent, it is also usual for specific deals and compromises to be made as part of the mediation process. For Karajarri, most categories of fresh water have been excluded by the way water is defined in the 2002 consent determination:

> 'the waters' means the waters within Determination Area A excluding flowing and subterranean waters;

> 'flowing and subterranean waters' means those waters within Determination Area A which are:

> (a) waters which flow, whether permanently, intermittently or occasionally, within:

> (i) any river, creek, stream or brook; and

> (ii) any natural collection of water into, through, or out of which a river, creek, stream or brook flows; and

> (b) waters from and including an underground water source, including water that percolates from the ground;

The same exclusions about water were made in the 2004 consent determination.

Such constraints leave native title holders with limited direct say in water decisions, thereby raising the importance of water licencing and planning processes in Western Australia to protect their native title rights and interests. Water planning is designed to provide for the domestic needs of native title holders (and others) before water is allocated for irrigation or other major use. Likewise the planning process is designed to ensure that sufficient water is retained in the natural environment to support ecological and cultural values including native title rights such as law, language, religion, identity, and living on country, including hunting and fishing. This planning approach recognises the key role of water in supporting valued relationships with healthy country.

Nearby to Karajarri country to the north-west, the Camballin irrigation project in the 1960s and 1970s is an instructive local example of the negative effects of poor water management and regulation. Here, the diversion and storage of surface water sources was undertaken through the construction of a dam, a creek and a barrage on the Fitzroy River, and land was cleared and levelled to grow crops such as sorghum and cotton. However, the project's aims were thwarted by monsoonal weather, and insect pests. This country was transformed by river channel erosion, the poisoning of wildlife from chemicals, and, as the project was abandoned, the land turning into a dust bowl (Yu, 1999: 10–11). Aware of such concerns, as part of their proposal WAI made the argument that they would be working with new irrigation technologies, that were less dependent on the heavy chemicals as these earlier Camballin cotton trials and other cotton trials in the Ord region of the Kimberley and they would not be growing cotton on a flood plain subject to flooding and erosion but on free draining soils with controlled trickle irrigation. Yet there remained many outstanding issues for Karajarri. As Karajarri man Mervyn Mulardy Jnr describes:

> They said they had a different technical style of irrigating cotton, but they wanted to use a lot of water. ... The cotton people were saying a lot of valuable water was being wasted out into the ocean. ... They [the Water and Rivers Commission] paid me to do consultant work between the traditional owners and them. The traditional owners were interested in other things that would not harm the environment, not cotton. Our old people said this water was not being wasted out into the ocean. (Mervyn Mulardy Jnr, interview, Bidyadanga, 14 October 2008)

Mervyn Mulardy Jnr, a co-author of this paper, was at the forefront of the campaign to stop the cotton, a campaign that extended to the Fitzroy River, and was also a leader for his community's battle for the recognition of their native title.

Karajarri had very limited native title fresh water rights, however, the WAI West Kimberley Irrigation Proposal did not go ahead. Karajarri were able to successfully oppose the cotton proponents with two key elements: native title rights to the *land*; and, the timing of administration processes in relation to the granting of water investigation licences.

In 1999, the Water and Rivers Commission had issued WAI with a groundwater investigation licence and required the company to go well inland from the coast to investigate and better understand the groundwater resource. To carry out this groundwater investigation work WAI needed land access to lands under Crown radical title which were also proposed for the recognition of Karajarri native title in the 2002 consent determination (described as area A). Hence before agreeing to this consent determination in 2002 WAI negotiated an Access Agreement with the Karajarri People which was included as Annexure A of the determination. The Access Agreement was specifically tied to the WAI's groundwater investigation licence from the Water and Rivers Commission. Unfortunately for WAI this licence expired in late 2004; they had neglected to renew it. Thus WAI's Access Agreement with the Karajarri became null and void.

To obtain a new groundwater investigation licence, WAI needed access to the land and permission from the land holder. However the 2002 Karajarri native title consent determination recognised Karajarri exclusive rights to area A, including 'the right to control access to, and activities conducted by others on the land'. Thus, in 2005 WAI were placed in the position of asking the opponents to the cotton proposal for land access for investigation work required before they could proceed with their cotton proposal. Because of this impasse, in 2005 the State Government allowed the Memorandum of Understanding between the State Government and WAI to expire without extension. The Water and Rivers Commission considered that WAI had not met their groundwater feasibility investigation obligations under the Memorandum of Understanding and were unlikely to do so and hence the Memorandum of Understanding should not be extended. WAI corporate leaders threatened to take legal action against the WA government for breach of contract, but they had no legal basis to do so.

The issues: three studies

Alongside this activity around cotton and native title, the Water and Rivers Commission began a water planning process to assess the proposal. The State government began with scoping the issues from talking with key stakeholders, and then setting out a work program for how these issues could be checked and investigated (Beckwith and Associates, 1999). Overall, three key studies were undertaken as part of the initial phase of the planning process: an overview of stakeholder issues by consulting company Beckwith and Associates (1999); a report on Aboriginal cultural values by anthropologist Sarah Yu (1999); and, a hydro-ecological survey of the wetlands of the north-western Great Sandy Desert by wetland scientists Vic and Christine Semeniuk (2000).

For the overview of stakeholder issues, Beckwith and Associates interviewed a diverse range of people, including Karajarri, Bidyadanga community people, public and private industry interests, relevant government departments, Indigenous and environmental organisations (such as the Kimberley Land Council and Environs Kimberley), and local government. Beckwith and Associates documented a range of general concerns made about the WAI proposal. Many stakeholders were concerned about the impacts of chemical pesticides and fertilisers that are used to grow cotton. People were most concerned about the possible contamination of the groundwater. Chemical exposure was also raised in relation to aerial spraying and the run-off of surface water after cyclonic events.

Stakeholders were also worried about the environmental impact of taking so much groundwater, and how this would affect domestic supply, pastoral operations, environmental features, and the possibility of future land uses. Beckwith and Associates also revealed that not much was known about the local Pindan ecology, and that this may result in the Pindan being disregarded as desert land. Further, this lack of knowledge about the Pindan extended to other key ecological features, including groundwater wetlands and the many unknown subterranean aquatic animals that likely live in the groundwater (stygofauna). Many people were concerned about the reduction in groundwater levels occurring without explicit knowledge of the effects of this loss (Beckwith and Associates, 1999: 4–6).

Additionally, Beckwith and Associates identified concerns about the impact of groundwater use on local stations, tourism operators, and the Bidyadanga community, with the key issues listed as:

- the potential negative impact on local area bores (stock and drinking water) through increased salinity or drawdown effects;
- the potential for pesticides and fertilisers to leach into the groundwater and affect local stock and drinking water resources; and
- the availability and suitability of groundwater for other existing or future agricultural activities in the area such as horticulture. (Beckwith and Associates, 1999: 7)

Some stakeholders were concerned that the Water and Rivers Commission would allocate all the groundwater, up to what was to be determined a sustainable limit, to the single development proponent WAI. The other issues that Beckwith and Associates identified were: the unclear relationship between native title and the allocation of water; and, the need to assess and protect local Aboriginal cultural values, particularly in relation to water, as well as traditional food sources more generally.

Because of the lack of knowledge about the groundwater and local ecologies, together with the scale of the proposal, many people emphasised the need to be cautious and careful in water planning through the application of the precautionary principle. The precautionary principle was devised by policy makers and ecologists to argue that 'where there are threats of serious or irreversible environmental damage, lack of full scientific certainty should not be used as a reason for postponing measures to prevent environmental degradation' (Ecologically Sustainable Development Steering Committee, 2002). Some stakeholders argued that the precautionary principle was important to apply up front, because it would be difficult both politically and administratively for the Water and Rivers Commission to stop or wind back an established cotton industry, if negative impacts were occurring.

As the main issues raised by the stakeholders included both environmental and cultural values, the Water and Rivers Commission contracted anthropologist Sarah Yu to lead a team working on Aboriginal cultural values, as well as a parallel study by wetland scientists Vic and Christine Semeniuk on hydro-ecological values (Yu, 1999; Semeniuk and Semeniuk, 2000).

The report prepared by Sarah Yu is based on interviews with Karajarri, Mangala, Nyikina, Yawuru, and Nyangumarta traditional owners, with a focus on Karajarri because most of their traditional country was within the study area, although, the significance of water stories is shared between all these groups. In the report, Yu details how water is described by the traditional owners in terms of life, and permanent water sources are known in Aboriginal English as *living waters*. Knowing where and how to get water is essential for survival in the Great Sandy Desert. Living waters may be surface waters, which may or may not require digging, and all are connected to the underlying water table. This underground water is called *kurtany,* which means 'mother of underground water' (Yu, 1999: 21). The living waters are all known as manifestations of *pukarrikarri* – the world-creation epoch of supernatural beings, also loosely described as the Dreaming. The living waters are evidence of the continuation of the *pukarrikarri*, and associated narratives, and are fundamental to understanding country, both geographically and temporally. Water sources can also be sites where Karajarri 'arose', as each Karajarri individual is intimately connected to sites in their country where they were created (Yu, 1999: 23). The living waters also relate to kinship and marriage arrangements (Yu, 1999: 33). Nyikina Elder Harry Watson has described these many relationships:

> There are inside stories for living waters, known all over the country. They've all got connections. The Law is big. It is not passive, it's active. We can't speak about this. It's not public. Water, culture and land. That's our *ngurrara* [country]. You can't divide them, doesn't matter which language you speak. (H Watson, interview at the Groundwater Committee meeting, Broome, 29 October 1999, in, Yu, 1999: 30)

Yu documents how Karajarri people also believe in *pulany,* water snakes or serpents, who reside in and, or, made the permanent water sources. These permanent water sources are called *jila* or *pajalpi*. Where there are *panyjin* reeds at the water sources, this often indicates the presence of *pulany*. *Pulany* have distinctive personalities, some are dangerous, others are docile, and active *pulany* travel underground, surfacing through escape holes called *tulkarru*. Most *pulany* interact with one another. Evidence of their activity is revealed in the formation of clouds, and storms with lightning and rain. When angry, the storms are violent, and the *pulany* can kill by drowning, battering or eating people (Yu, 1999: 18–20). Because the *pulany* are unpredictable, Karajarri do not camp in the immediate vicinity of living waters.

The changing seasons are 'put down by *pukarrikarri*' and the seasonal replenishment of rain is promoted by the *pulany* who make the living waters such as *jila*. Karajarri understand the health of the living waters as an indicator of the health of country. They must never dry up. When the living waters fill up, the other water sources in country will be replenished. Thus, *kurtany,* the mother of the ground water, through the generation of rain by the *jila*, makes everything strong and healthy. With the replenished springs, the trees, animals, plants and people live off this water until the next rains. Even saltwater fish, such as mullet and bream, drink the fresh rainwater during the wet by coming to the surface of the ocean with their mouths open (Yu, 1999: 29–30). In this way the living waters are not just the permanent springs, but generate life for the whole of country, people included.

It is clear from these interviews that Karajarri must respect the seasonal cycle of water use and replenishment for their own survival, and this is also their responsibility as traditional owners. This respect will ensure that the water table or water level is maintained in the permanent sources of water. There are correct protocols, and knowledge passed down through song and narrative to ensure future generations will be able to continue these responsibilities. There are rainmaking ceremonies to persuade the *pulany* to 'get up' and make rain. Also, *pulany* can identify strangers and Karajarri were careful on fieldwork to ensure such people were appropriately introduced to the water sources. A Karajarri Elder, since passed on, spoke about the importance of this work:

> We have to look after this water. If the water go, everything will be finished. Life gone. Spirit gone. People gone. The country will have no meaning. (JDN, interview, Bidyadanga, 8 October 1999, in, Yu, 1999: 32)

Wetland scientists Vic and Christine Semeniuk had as their objective the identification of wetland types, especially those wetlands of significance, and the determination of the hydrological mechanisms maintaining them. The wetlands were assessed as to their 'value' and 'function': value was described as the importance, merit, or worth of a wetland after evaluation; and function being the role a wetland performs in its natural setting (such as water and food source, habitat, drought refuge, breeding ground, collection point for a range of ephemeral drainage lines, and hydrological discharge zone). Semeniuk and Semeniuk categorised and described different wetland types, and the assemblage of plants that occurred in each habitat. They found that the wetlands are maintained in four ways: surface water flows from drainage lines; water table rise; the ponding of water by the near surface hardpans; and, by upward leakage from formational waters (formational waters are waters that were either originally rainwater or sea water, but have since undergone substantial transformation). Their report divided the hydrology of the Great Sandy Desert into three zones: the northern section where mainly fresh groundwater resides at depth in the Broome Sandstone, with the water table in excess of ten metres below the surface; a southern section where fresh, brackish to saline groundwater is shallow, located beneath a calcrete and limestone sheet; and a coastal section, where fresh to saline groundwater is found in coastal aquifers. They described the Broome Sandstone as not internally homogenous in terms of rock types and cementation, with the transmission of water being a complex relationship in different sedimentary environments; sand dominated, to fluvial (sand and mud), and coastal tidal flat (sand with more mud sheets).

Along the coast, the variety of wetlands are categorised according to whether the wetlands are a combination of fresh and salt water (such as the mangrove systems), or where impeded freshwater has formed small to large clusters of lake-like wetlands. There are also springs along the upper shore zone and low tidal zone where freshwater has found pathways through the Broome Sandstone. With inland wetlands, Semeniuk and Semeniuk categorised a range of types: those that have developed along drainage lines with large fluvial discharge, these wetlands are described as 'windows to the regional water table'; those that are also located along drainage lines, but have overflow whereby water ponds to form clay-lined pans; clay or muddy sand floored wetlands that are along relict drainage lines; pans that have developed along low relief drainage lines; and peat mounds, which are wetlands formed around springs. Particular areas challenge these neat categories, such as the Salt Creek Line which has a range of sediment types and wetlands.

Semeniuk and Semeniuk concluded that the wetlands are varied and complex in origin and maintenance, with significant values in terms of biota and geo-heritage. Further, that the Broome Sandstone cannot be categorised as a simple unconfined aquifer. Because of the complex pathways and springs of inland and coastal wetlands, Semeniuk and Semeniuk recommended that the dynamics of the groundwater be further investigated to better understand how these wetlands are maintained. They advised against assuming that abstracting water from the Broome Sandstone would not affect the wetlands in the region. This advice contradicted the argument that the fresh groundwater was being wasted out to sea. The groundwater was shown to support culturally and ecologically significant wetlands along the coast and inland.

Together, the three reports by Beckwith and Associates (1999), Yu (1999) and Semeniuk and Semeniuk (2000) revealed what an enormous impact a single large development proposal could have on the local community, local futures, traditional owner identity and relationships with country, and the health of complex freshwater ecosystems in a very arid environment. Karajarri most directly expressed how water is a life-giving source, and a key relationship that they must respect to ensure the health of country. However, this relationship was also central to the two reports on stakeholders and hydro-ecological values, which revealed the many activities and connections that have thrived on a healthy water source. Rather than treating water as just another resource input for the irrigation industry, the reports illustrated how the extraction and consumption of large amounts of water could widely compromise local reliance on La Grange groundwater as a life source.

The process of engagement

The engagement between Karajarri and the Water and Rivers Commission is one of Australia's first examples of a substantial consultation process with Indigenous people over a specific water allocation (Jackson, 2009: 43). There were three key parts to this consultation:

- the methodology for the cultural and ecological studies;
- the focus on process issues around stakeholder involvement; and
- the recognition of the distinct role of traditional owners, especially Karajarri;

A significant innovation in the engagement was that the Water and Rivers Commission instructed the wetland scientists Semeniuk and Semeniuk and the anthropological team led by Sarah Yu to undertake their fieldwork together. Through this experience, the traditional owners and the scientists enjoyed exchanges in their expertise. The wetland scientists gained local knowledge of the location of wetlands, as well as their water permanence throughout the seasons and over the many years of Karajarri observation and knowledge. This helped Semeniuk and Semeniuk to work more effectively within their limited budget and time. Karajarri were able to ensure the scientists' work was conducted according to cultural protocols. Karajarri and other traditional owners were also very interested in the scientific understanding of groundwater hydrology and how the wetlands were supported. Mervyn Mulardy Jnr talked about the exchange:

> Bidyadanga water is very old, maybe over a million years old. We went to a place called Juwurr-kara and Juwurr-kakara jilla springs. A lot of places had Baler shells, which the old people used to carry water. The scientist checked this waterhole and got a shock. In the whole of Australia, the purest water is the monsoon rains in north Queensland. But he was fascinated that he found it at the same level as that waterhole. He couldn't get over it. (Mervyn Mulardy Jnr, interview, Bidyadanga, 14 October 2008)

The two teams also found they had much commonality in their knowledge, as Mulardy Jnr said:

The old people had names for different levels of ground water, and how it moves. At the end of the day, everything the old people were saying matched up with the scientific version. So the scientific knowledge and traditional knowledge were the same. (Mervyn Mulardy Jnr, interview, Bidyadanga, 14 October 2008)

Both the traditional owners and the wetland scientists agreed on many points, and this was reported in both reports. There was agreement on the direction of groundwater flows at particular sites, the layers of groundwater, and that the coastal springs are fed by underground water that comes from the high country inland (Yu, 1999: 34). There was also agreement across the categorisation of different types of wetlands (see Tables 5.2 and 5.3). However, even though there is a clear link between cultural values and water places, cultural knowledge and ecological knowledge cannot be equated (see Weir, this volume). There remains a strong imperative for the role of Indigenous people and their knowledge in water decision making, which cannot be replaced by scientific expertise.

Table 5.2: Terminology employed by local Aboriginal people for the wetlands

Water type	Description
lirri	Soaks, in which water is dug up for drinking. Some are permanent (lakes), others are dug up in the hot time (sumplands and damplands).
jila	Permanent water sources. In some cases *jila* have visible surface water (lakes and peat mounds), but many require digging. A *jila* may be marked by a small depression in the ground. Scrubby t-trees may surround the water. *Jila* occur in clayey soil from which the white mud *kalji* is found.
pajalpi	Ecosystems surround springs, as permanent water sources found on fringes of coastal mudflats, or inland areas.
wawajangka	Fresh water seepages found in mudflats in the intertidal zone and only accessible at low tides.
pirapi	Claypans (sumplands) that fill with water after rain, and usually dry up after the rain or as the hot time approaches.

Source: Semeniuk and Semeniuk (1999): 12. The terms in brackets signify the classification that was used by the scientists.

Table 5.3: Practical classification, technical classification, and Aboriginal terminology

Practical classification	Technical classification	Aboriginal terminology
oasis	lakes, sumplands, playas	*jila,* some *pajalpi*
pan	sumplands, damplands, playas	*lirri, pirapi*
spring	peat mounds	*pajalpi*

Source: Semeniuk and Semeniuk (1999): 13.

The interactive fieldwork methodology also provided additional opportunities for dialogue about the many close connecting relationships with ecological and cultural values. The close relationships held are reflected in the maps the teams produced, which showed a strong correlation between the ecological and cultural sites. Key areas included the wetlands that form a parallel line just inland from the coast, and the Mandora Marshes and Salt Creek System that include one of the furthest inland saltwater mangroves in Australia. Overall, the maps of the study area reveal that the cultural sites are more widespread than those sites of high ecological value. This mapping work enabled the Water and Rivers Commission to undertake their own preliminary mapping to guide as to where possible irrigation development might or might not occur. This mapping work is reflected in the management zones in the most recent La Grange management plan (Figure 5.1). However, even though there is a clear link between cultural values and water places, cultural knowledge and ecological knowledge cannot be equated (see Weir, this volume). The role of Indigenous people and their knowledge in water decision making cannot be replaced by scientific expertise.

A key objective of the report on stakeholder issues was to identify stakeholders and a process for further community involvement (Beckwith and Associates, 1999: 3). This reflected the priority of the Water and Rivers Commission, who wished to identify the key people for ongoing dialogue about the West Kimberley Irrigation Proposal, its development and sustainability, as well as how these people would be involved in the water planning process.

Out of the scoping work undertaken by Beckwith and Associates, all the key stakeholders agreed upon how the Water and Rivers Commission would consult in developing the allocation plan (Figure 5.3). The La Grange Groundwater Committee was formed in 1999 with representation from 18 key locally based stakeholders (government and non-government) with a significant interest in water issues. The Committee, which had an independent chairperson, provided comment and advice to the Water and Rivers Commission on water allocation issues associated with the La Grange groundwater area but did not have any formal decision making powers. Separate dialogues were held for local users, other stakeholders and the general public. Karajarri and the Bidyadanga Community Council had representation on the Groundwater Committee to present their issues and perspectives.

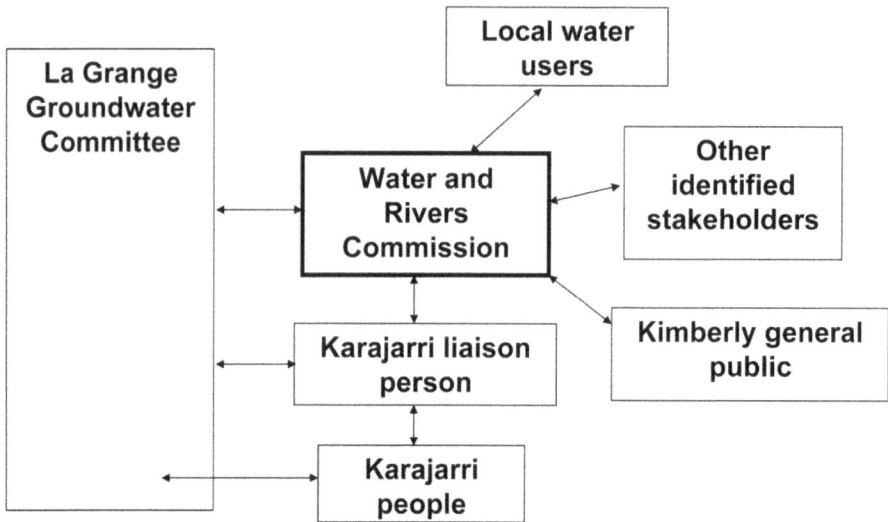

Figure 5.3: Public involvement model for the La Grange groundwater allocation planning process.

The Water and Rivers Commission also appreciated the importance of direct dialogue with Karajarri as the main traditional owners for the planning area they had identified for the La Grange groundwater. Protocols were established to ensure this relationship was respected. In Bidyadanga, the Water and Rivers Commission would meet first with the Karajarri, before meeting with the Bidyadanga Community Council. This enabled direct discussion of the upcoming meeting agenda, the exchange of concerns, and to update the Commission of any Karajarri news. By doing so, the Commission were able to communicate with Karajarri as a group, rather than just the Karajarri representatives on the Groundwater Committee. Further, Mervyn Mulardy Jnr was employed as the Aboriginal Resource Person for La Grange groundwater planning, a liaison position to provide the interface and translation between the Commission and Karajarri.

After the failure of the cotton proposal, the new authority, replacing the Water and Rivers Commission, the WA Department of Water re-commenced the planning process for La Grange water. The 'Draft La Grange groundwater sub area water management plan – allocation' was released in late 2008 for public comment. As part of this planning process, innovative engagement with the traditional owners has continued. The Department of Water funded a film in plain English and Karajarri to describe the new consultation process and water allocation plan (Mathews, 2008). The film begins with the camera zooming into the *panyjin* reeds at a *jila*, which are often the indicators of *pulany* being present. Mervyn Mulardy Jnr speaks throughout the film about how the planning process is to share the water 'right way', with people from other communities, farms, cattle stations, and outstations, and to look after the water to keep it healthy and alive. He explains that the government will take some La Grange groundwater every year, and that Karajarri will need to make sure there is enough water for their culture and future development. Mulardy Jnr describes the water licences, the paperwork, and how licences do not give ownership to the water, and the ongoing government process of monitoring those licences. The film was made to encourage people to engage with the process because, as Mulardy Jnr says in the film, 'It's our right to look after all *ngapa* (water) places, and we've got to make the right decision.'

The active engagement approach by the Department of Water with Karajarri was also evident in a recent licence application from the pastoralist managing Nita Downs station, who wished to diversify into horticulture. The Department arranged for a consultation meeting on Nita Downs between Karajarri and the station leaseholders about the proposal, to ensure cultural and other Indigenous interests were allowed for as part of the licence application assessment process, as well as to better inform Karajarri about water licences and regulation. During the trip, Karajarri took the opportunity to harvest bush tucker. Karajarri man Joe Edgar has spoken about how the pastoralist was astounded by this food collection, which he did not realise was so readily available (Edgar, 2009).

Such levels of engagement are respectful of the issues being raised by the local community, and the unique relationships traditional owners hold with country. It has set a standard for future engagement with Indigenous people in water planning in Western Australia and Australia. Although the Water and Rivers Commission have been criticised for not engaging more broadly with traditional owners, including Yawuru people to the North and Nyangumarta people to the south, as the La Grange groundwater involves a much bigger area than Karajarri country.

The National Water Initiative requires Indigenous people to be involved in water planning, and their issues to be considered in water management, however in WA there is no such state legislation enforcing this. There was no legislative or policy need for the Department of Water to facilitate good engagement processes with Karajarri, such as the Nita Downs trip. Despite all this good work, these processes are not securely imbedded in law and policy. Generally, in WA there is a lack of Indigenous peoples' involvement in water resource monitoring and management, a lack of policy on Indigenous peoples' access to water and economic use, and a lack of adequate equitable water service provision for many Indigenous communities.

Water decision making

The original planning around La Grange water allocations took place in a context that was dominated by a single large scale cotton irrigation proposal. In 2010 the Department of Water finalised the La Grange water allocation plan. The plan has rules for water use and the monitoring of that use which give consideration to ecological, social and cultural values. The plan sets the allocation of water for consumptive purposes at about a quarter of the initial estimate of water potentially available when the WAI cotton irrigation proposal commenced. A total allocation limit of 50 gigalitres per year is established for all consumptive uses in the La Grange Basin (Department of Water, 2010: 37). Water is still allocated on a first come, first allocated basis, which will advantage those development proponents who have a proposal ready to apply. The Department of Water acknowledges that the traditional owners require certainty that a commercial water allocation will be available if and when they seek it, and is considering how to address this (Department of Water, 2010: 19). Karajarri still worry if the water plan will be 'right way' for country (Mulardy Jnr cited in Mathews, 2008). There is always the opportunity for an agreement between the government and traditional owners to facilitate their greater involvement in water planning and allocation, although this is constrained by narrow interpretations of native title.

The problem with planning is that it is hard to evaluate activities before they are undertaken. This is why the precautionary principle approach is so important. The La Grange allocation plan acknowledges the complexity of the salt and fresh water interface and the importance of increased management and monitoring before any significant increase in water use occurs. As Mervyn Jnr said in the film, the Plan has 'harder rules' for the significant ecological and cultural areas along the coast, and the Mandora Marshes and Salt Creek System. Applicants for new water extractions are required to demonstrate that their proposed development will be sustainable in the long term. The Department may require additional work to be done by the applicant to make sure that groundwater extraction has negligible impact on the social, cultural or ecological values of wetlands or other groundwater dependant ecosystems.

The Karajarri's native title determination over most of the area of the La Grange Basin makes them a major player in any proposed large scale groundwater development. In determination area A, which covers Frazier Downs and a substantial area of Crown radical title, Karajarri people have exclusive native title rights. This includes 'the right to control access to, and activities conducted by others on the land'. As with the WAI proposal the Department of Water would require any applicant for a large scale development to carry out regional level groundwater investigation work on the inland Crown radical title to be able to determine groundwater availability. The applicant would need to negotiate an access agreement with the Karajarri to do this work. The Karajarri also have exclusive native title rights to the Frazier Downs pastoral lease area and could develop a sustainable irrigated agriculture business on this land.

Overall, three key studies were undertaken as part of the initial phase of the planning process: an overview of stakeholder issues by consulting company Beckwith and Associates (1999); a report on Aboriginal cultural values by anthropologist Sarah Yu (1999); and, a hydro-ecological survey of the wetlands of the north-western Great Sandy Desert by wetland scientists Vic and Christine Semeniuk (2000).

Concluding remarks

Water issues bring people together through the far reaching connections that water sustains. The life sustaining properties of water were acknowledged in the three reports by Semeniuk and Semeniuk (2000), Beckwith and Associates (1999) and Yu (1999), albeit in different languages with different priorities, values and methodologies emphasised. The water management and planning work of the Department of Water uses another language again, but also emphasises protecting groundwater dependant cultural and ecological values and managing sustainable use. Where people share commitment to sustaining fresh water ecosystems, their different knowledge traditions, skills and approaches generate a dynamic dialogue for water planners and decision makers to draw on.

The WAI large scale cotton proposal focused on treating water as a commercial resource and was not sensitive to considering the other roles of water, including being a fundamental life source. The proposal ultimately failed. Whilst this opportunity for development was forgone, Karajarri and many others are able to continue to enjoy all the life of the desert country that is sustained by La Grange groundwater. They also have the opportunity to plan for and invest in development proposals that may have stronger links with and outcomes for the local community. Critically, the failure of the WAI proposal has enabled a much longer time frame for water planning, research and engagement with key parties.

What is unique about the La Grange planning work is how the joint cultural and ecological fieldwork revealed the common ground held between Indigenous and hydro-ecological knowledge, thereby facilitating the immediate relevance of Karajarri water knowledge to water management and planning processes. Karajarri water knowledge is multi-layered, complex and restricted, and central to their native title. This is productive knowledge that can be engaged with in water management and the development of sustainable water consumption industries. Good process is central to ensuring this engagement is meaningful.

Negotiations over water will intensify as proposals to develop agriculture in northern Australia gain traction. Aboriginal people are central to these negotiations, as evident in the recognition of their native title rights and their extensive knowledge about country. Whilst, as yet, Karajarri do not have recognised rights to a water allocation, they clearly wish to be involved in water management and planning. The time and energy invested into this water management planning for the La Grange groundwater, has hopefully provided for a water management plan which all parties can identify with. In any case, Karajarri will continue to work to ensure water will be there for country, for future generations of native title holders, and for everybody else, because the living waters are central to Karajarri people, law, country and their way of life.

References

Beckwith and Associates 1999, 'La Grange Groundwater Allocation, A Kimberley Sub-Regional Allocation Plan, Overview of Stakeholder Issues', report prepared for the Water and Rivers Commission, Perth.

Department of Water 2010, 'La Grange Groundwater Allocation Plan, Water Resource Allocation and Planning Series', Report no 25, Department of Water, Perth.

Department of Water 2008, 'Draft La Grange Groundwater Sub Area Water Management Plan – Allocation', Department of Water, Perth.

Ecologically Sustainable Development Steering Committee 2002, *National Strategy for Ecologically Sustainable Development*, part 1, Department of Environment, Canberra.

Edgar, J forthcoming, 'Indigenous Land Use Agreement – Building relationships between Karajarri traditional owners, the Bidyadanga Aboriginal Community, La Grange Inc and the Government of Western Australia', *Australian Aboriginal Studies*.

Edgar, J 2009, presentation to the National Indigenous Water Planning Forum on 19 February 2009, Adelaide, National Water Commission.

Jackson, S 2009, *National Indigenous Water Planning Forum: Background Paper on Indigenous Participation in Water Planning and Access to Water*, CSIRO, Darwin.

Mathews, D 2008, *Ngapa,* translation DVD for the La Grange Water plan, English, Kriol and Karajarri language, Department of Water.

National Native Title Tribunal 2009, *Determination of Native Title: Karrajarri People A (2002) Karajarri People B (2004)*, Geospatial services, National Native Title Tribunal.

— 2011, *Kimberley Native Title Applications and Determinations Areas*, Geospatial services, National Native Title Tribunal.

Ross, J 2009, 'Sustainable Development in Northern Australia', A report to Government from the Northern Australia Land and Water Taskforce, Department of Infrastructure, Transport, Regional Development and Local Government.

Semeniuk V and C Semeniuk 2000, 'Wetlands of the Northwestern Great Sandy Desert in the La Grange Hydrological Sub-basin', report prepared for the Water and Rivers Commission, Perth.

Strelein, L and J Weir 2008, 'Our Public and Private Relationships with Water and Native Title', seminar presentation, Australian Institute of Aboriginal and Torres Strait Islander Studies Seminar Series, 25 August 2008.

Weir, JK 2011, *Karajarri: A West Kimberley Experience in Managing Native Title*, Discussion Paper Number 30, Native Title Research Unit, Australian Institute of Aboriginal and Torres Strait Islander Studies, Canberra.

Yu, S 1998, 'Land Interests of Karajarri in the Area around Port Smith Caravan Park', report to support application to the Indigenous Land Corporation to purchase special lease 3116/9994: Port Smith Caravan Park.

— 1999, 'Ngapa Kunangkul: Living Water', report on the Indigenous cultural values of groundwater in the La Grange sub-basin, prepared for the Water and Rivers Commission, University of Western Australia, Perth.

Cases

Mabo v Queensland [No 2] (1992) 175 CLR 1

Nangkiriny v Western Australia [2002] FCA 660 (12 February 2002)

Nangkiriny v Western Australia [2004] FCA 1156 (8 September 2004)

Western Australia v Ward [2002] 213 CLR 1 (8 August 2002)

Yanner v Eaton [1999] 201 CLR 351 (7 October 1999)

Legislation

Native Title Act 1993 (Cth)

6. Native Title and Ecology: Agreement-making in an Era of Market Environmentalism

Lee Godden

Agreement-making, given particular impetus by the advent of native title, forms an important component in establishing a stronger presence for Indigenous peoples in ecological protection and environmental management (Tehan *et al.*, 2006: 1–2). Aboriginal and Torres Strait Islanders' customary care for country clearly continues apart from such formal western structures, but of necessity must interface with non- systems (*Yanner v Eaton*, 1999: 76). Predominately, the settler institutions for environmental protection have been built upon an ecological perspective, but these structures and values systems are being substantially reworked through the increasing influence of market environmentalism. Historically, settler models for managing ecosystems have struggled to provide meaningful and effective participation by Indigenous peoples (Hill and Williams, 2009: 161); and while strong potential exists for market environmentalism to offer avenues for more robust participation for Indigenous peoples that is yet to be effectively realised. Accordingly, this chapter explores the dynamic of agreement making in the native title/ecology sphere, and considers the challenges and opportunities posed by market environmentalism for the protection and management of Indigenous peoples' communally-held land and resources and corresponding economic and cultural sustainability. Specifically, it examines how ecology and native title are co-located within a legal, economic and social space framed by the particular contractual/exchange relationships based upon negotiated agreements.[1] This 'space' between native title and ecology is increasingly perceived as operating within wider structural changes precipitated by globalisation, public/private partnerships and market mechanisms: elements of which are all apparent in market environmentalism. In earlier periods, major structural change has marginalised many Aboriginal and Torres Strait Islander peoples excluding them from a range of social and economic benefits, including labour force and industry participation, and precluded due recognition of Indigenous knowledge and customary practices for the sustainable management of country.

---

1 Agreements are not confined to the ambit of the *Native Title Act 1993* (Cth) and state counterpart legislation. However many agreements arise 'in the shadow' of the legislation.

Connection with country: competing paradigms of environmental governance

Within Australia, the recognition of native title in 1992 by the common law gave specific legal force to Indigenous peoples' involvement in the protection and management of country (*Mabo v Queensland [No 2]*, 1992). While Aboriginal and Torres Strait Islander peoples have long cared for country under law, tradition and custom, *Mabo No 2* gave belated acknowledgment in the settler legal system to the significance of the relationship between Indigenous people and their governance of land and waters, through the concept of connection.[2] Despite the limitations of the construct of native title in capturing the dynamic of Indigenous peoples' relationship, such legal recognition through common law native title, and the subsequent enactment of the *Native Title Act 1993* (Cth) gave Indigenous people, 'a seat at the table' in terms of seeking legal protection for their rights to care for country (Strelein, 1993: 38–39). Since 1992, there has been considerable expansion in the areas in which Indigenous peoples have gained responsibility under the settler legal system for environmental protection and management, with a growing number of determinations of native title.[3] Further, much of the expansion of Indigenous rights to care for country beyond formal native title determinations has been achieved through the instigation of agreements, either directly under the framework for Indigenous land use agreements under the *Native Title Act 1993* (Cth)[4] or pursuant to broader agreement making[5] – much of which was initiated consequent to the legal recognition of native title. The expansion of Indigenous protected areas across northern Australia is one such example where a principally 'environmental' regime has expanded to offer a more receptive forum for Indigenous care for country.

The increasing prominence of native title determinations and agreement-making in the sphere of environmental protection and management occurred in the context of a growing recognition within western knowledge systems of the importance of holistic and integrated ecological understanding of land and waters. Ecology, itself a challenge to the reductionism of many western scientific disciplines, is predicated upon a systemic approach that considers the interconnections and interrelationships between elements in the environment – including people. Ecology, as a guiding paradigm for environmental and

---

2   For legal requirements to establish connection, see *Native Title Act 1993* (Cth): s 223 and as elaborated by relevant case law; primarily *Members of the Yorta Yorta Aboriginal Community v Victoria* (2002) 214 CLR 422.
3   Essentially there are two forms of 'determinations' of native title under statute – one the result of a litigation process and the other where there is a consent determination *Native Title Act 1993* (Cth): s 225.
4   *Native Title Act 1993* (Cth) Part 2 Div 3 Subdivs B, C, D and E. On the relationship between ILUAs and other 'future acts', note s 24AB(1).
5   For an overview of agreements see the Agreements, Treaties and Negotiated Settlements Database at: <http://www.atns.net.au/> (accessed 9 May 2009).

natural resource management, tends to emphasise the integrity and uniqueness of natural systems, and accordingly the need to preserve such systems apart from 'human interference'.

Over the last decade or so, the two constructs of native title and ecology have formed major points of institutional and organisational structure around which agreement making has focused with respect to customary care for country. Initially, there were particular points of tension between ecological understandings of wilderness and ecological integrity, and Indigenous approaches which emphasised the integral cultural connection and relationship between Aboriginal and Torres Strait Islander people and their traditional country. More recently, there is a greater appreciation of the inherent tie between Indigenous cultural identity, and connections to land and waters within a prevailing western ecological conservation paradigm (DEH, 2001). Although the trajectory is not always smooth, as tensions around the *Wild Rivers Act 2005* (Qld) in northern Queensland can attest. Moreover, the need for more inclusive Indigenous participation in environmental protection is widely acknowledged, if not always achieved in a truly collaborative manner given power asymmetries at play in many formalised modes of environmental law and management that seek to adopt deliberative or collaborative models, (Hill and Williams, 2009: 163, 168). Agreement-making has played a major role in reorienting the understanding of the dynamic between native title and ecology to give greater prominence for Indigenous peoples. This interaction between native title and ecology, which has been growing in significance, has been built upon particular assumptions of the role for western science and traditional knowledge systems (Verran, 2008). These approaches, and the interface between them, currently inform much of the existing administrative and governance structures for environmental protection. Such institutional arrangements and disciplinary paradigms arrangements are now being challenged in various ways by the rise of market environmentalism.

Market environmentalism, denoting a complex of regulatory, structural economic social, cultural and institutional changes has assumed an increasing role in natural resource management and environmental protection in Australia over the last decades (Eckersley, 1995: 7). Presently, it is being actively promoted through the idea of environmental sustainability. These influences, fashioned by both global and local factors have reshaped many aspects of the interface between Indigenous peoples' communally held land and resources, and western modes of environmental management across many countries. In turn, these changes have the potential to reorient the understanding of key concepts in Australian law and policy, such as native title and ecology, and the dynamics of the interaction between these spheres. Accordingly this chapter seeks to critically interrogate the nature of the changes occurring under the rubric of market environmentalism to probe whether these policies do offer sustainable long term outcomes for Indigenous communities in their relationship with 'country'. In particular, the chapter investigates pressures associated with market environmentalism to individuate communal land and resource holding associated with native title and to introduce western forms of property and contractual-based environmental regulation and natural resource regimes. While new opportunities may arise for Indigenous economic empowerment through environmental commercialism in fields such as ecosystem services, it is necessary to evaluate whether these agreement and exchange based relationships do offer a compelling means to achieve both community empowerment including appropriate recognition of customary values and knowledge and self determination, as well as effective and responsive care for country in the native title context.

From ecology to native title

Ecology

Over the course of the second half of the twentieth century, recognition of the importance of environmental protection became entrenched in many societies such that environmentalism is now 'as much a state of being as a mode of conduct or a set of policies ... [c]ertainly it can no longer be identified simply with the desire to protect ecosystems or conserve resources' (O'Riordan, 1981: ix). Ecology, the new integrative science emerging in the 1960s, and closely identified with environmentalism, illuminated the need for law, policy and management practices designed to arrest the rapid decline of ecosystems that had been precipitated principally through industrialisation, urbanisation and colonisation. Responsibility to arrest the decline was predominately seen as resting with nation state institutions. The 1970s saw a rapid proliferation in the international law and administrative structures, such as the United Nations Environment Program, which sought to protect the natural environment (Fisher, 1999: 372). Developments at an international level had parallels in most western democracies, including Australia, which saw the first comprehensive platform of environmental legislation introduced in the mid 1970s. Since then a comprehensive legislative and institutional framework has been implemented with extensive federal government and state government involvement principally through departments of environment but with significant institutional responsibilities for land and water management within many other government departments, including market regulatory authorities (Godden and Peel, 2010: 125). Ecologically sustainable development, since its policy genesis in 1992, has remained the primary guiding principle informing regulatory objectives under most environmental legislation (Peel, 2008).

In turn, social values have shifted over time, with conceptual notions of what constitutes 'the environment' being fluid as demonstrated by current debates over how water should be perceived against a background of predicted climate change and highly variable precipitation patterns in Australia. For example, should water be seen as a resource, a fundamental component of an ecosystem, a commercial tradeable entity, an ecosystem service, a human right or as a cultural value and native title right? While the parameters of the concept of environment, and indeed natural resources, has shifted under prevailing discourses, the present phase is characterised by growing ascendancy of the neo-liberal paradigm of market mechanisms and deregulatory approaches (Kinrade, 1995: 86). In this regard, such perspectives offer an at times competing, and at other times, a congruent approach to the previously dominant concept of ecology that gave precedence to western scientific knowledge as the basis for environmental governance, management structures and institutions.

Indeed, since at least the seventeenth century, western scientific discourses about nature have played a formative, often decisive, role in determining notions of the environment that have been promulgated internationally. This discourse was highly influential in the expansion of western environmentalism into other cultural contexts; typically, often, as a consequence of colonisation (West *et al.*, 2006: 251). Environmentalism has on the one hand contributed to the post colonial construct of development (Blaser *et al.*, 2004: 3), while simultaneously criticising the dominant 'growth' ethic of development regimes. Thus, while in the twentieth century the emergence of the science of ecology was a significant catalyst for the invigoration of holistic approaches to nature (for an overview of the rise of ecology, see, Worster, 1994), it was not unproblematic, at the very least, in its interface with indigenous cultures. The inception and growth of the global conservation movement and consequent creation of concepts of 'wilderness' and 'ecological integrity', integral to western scientific conceptions of ecology distinguish 'nature' and 'culture' as antitheses. Such a distinction is significantly different to Indigenous understandings of country (see, for example, Hokari, 2005: 214–222). Further, many of the institutions and laws pertaining to what might be termed 'first phase' environmentalism remain essentially grounded in this distinction. Such a distinction removes agency from the environment; instead constituting it as a space to be protected or plundered (Massey, 2005: 86; Kinnane, 2005: 195–222). Such categorisation has not only removed agency from the environment itself, but also from Indigenous peoples who have historically inhabited and managed country, with consequent social, economic and cultural ramifications for Indigenous communities in many parts of Australia.

Ecological understandings of the environment later became framed predominately as 'biodiversity'. International developments, particularly the influence exerted by international legal instruments, such the Biodiversity Convention in 1992 (United Nations Environment Program, 1994), were highly influential in the further translation of key constructs to regulate and protect the natural environment. This emphasis signified a shift from a focus on the unique and special 'bits' of nature to a more pervasive approach that sought to achieve diversity and complexity in all natural systems rather than selective protection. This influence is clear in national environmental law frameworks, where the rise of ecologically sustainable development was a major trend from 1992 (Ecologically Sustainable Development Steering Committee, 1992). The emergence of these ideas of sustainability coincided with the belated recognition of native title. Thus ecology, with its emphasis upon integration and interdependency had the potential to re-situate human beings as a component of a wider system of web-like interactions involving myriad physical and non-physical elements (Dovers and Price, 2007: 36). Yet despite the broadening of many legal definitions of 'environment' that occurred (see, for example, *R v Murphy*, 1990), the practice of 'separate' treatment of the natural and cultural world largely continued in many spheres of environmental law and management and it remains a guiding assumption in many key natural resource management fields.

In light of the trajectory that developed for the preservation and conservation of ecology within Australia from the 1980s, the culturally-imbued understanding of country that characterises Indigenous relationships to 'ecology' was accorded little direct acknowledgment in key legal and policy frameworks for the conservation and protection of the environment (Plumwood, 2003: 51). Cultural heritage concepts continued to play a significant role in providing some avenues to include Indigenous peoples' perspectives of the value and protection to be accorded to the natural environment, although the limitations of the early cultural heritage frameworks have been widely recognised (see, for example, Thorley, 2002: 110). More holistic framings of Indigenous cultural heritage are now apparent, but these concepts are rarely integrated fully into environmental protection laws.

Thus at least initially, many environmental laws and policies had the effect of excluding Indigenous peoples from participation in mainstream environmental and natural resource management (NRM). Recently, a more dynamic understanding of the complexities of environmental impacts and their social consequences has emerged, consistent with a growing recognition of the significant role of socio-cultural factors in environmental conservation and NRM (see, generally, Langton *et al.*, 2006). Such approaches, potentially, are more open to Indigenous knowledge and values in the care for country. Indigenous cultural identity is intimately bound with environmental conservation, and should be recognised within legislative and policy regimes which are open to that which Langton describes as Indigenous 'life-ways', as a part of a broader reconciliation project (Langton, 2003: 142). Agreement-making and recognition of native title have performed an important function in 'opening up' environmental protection regimes to Indigenous 'lifeways'.

Australia also has been influenced by the trends at a global level towards greater involvement of Indigenous peoples in managing areas for ecological protection although the extent of 'partnership' and the degree of autonomy accorded to Indigenous peoples in environmental governance and management varies widely. Governance models for implementation of Indigenous involvement in land and water management are diverse, traversing a spectrum from mere consultation to direct decision making (for an overview see Nettheim *et al.*, 2002). Thus while there have been strong calls to create more participatory frameworks for Aboriginal and Torres Strait Islander people, many trends to involve Indigenous peoples within mainstream environmental and NRM management regimes have been criticised as occurring within an assimilation framework. Such schemes have been regarded by some as, predominately building the capacity of Indigenous communities to successfully operate within post colonial environmental management regimes without delivering sustainability and self determination for Indigenous communities (Strelein, 2004: 189, 196–198). Thus where concepts of management are still predominately grounded in the paradigms of western environmental conservation including market environmentalism, meaningful incorporation of an Indigenous worldview may be limited. Accordingly, in approaching environmental conservation with Indigenous communities, concepts of value and significance must be open to Indigenous perceptions and notions of responsibility for a more effective incorporation of Indigenous peoples' care for country. As Ross and Ward note,

> modern land management requires reversing degradation at accepting the concept of 'peopled landscapes' as a fundamental and essential part of a healthy and sustainable environment. Therefore the knowledge values and perspectives of local Indigenous people are now seen by progressive natural resource managers as vital to achieving a more comprehensive and holistic approach to land management … because approaches based on Western science alone have so clearly failed. (Ross and Ward, 2009: 37–38)

Yet despite many acknowledgements of the need for greater recognition to be accorded to Indigenous value systems in ecological and natural resource management, there are surprisingly few successful examples of inclusion of indigenous engagement in natural resource management. Hill and Williams (2009: 172) after a detailed case study of how indigenous marginalisation occurs at a local or project level, propose a number of suggestions to build more inclusive national level policy responses. A key initiative that is proposed is to establish a separate funding program for Indigenous NRM and recognising the role of Indigenous NGOs who focus on environmental governance and NRM as vital to success.

Thus one of the most marked shifts in delivering more 'on the ground' autonomy for Indigenous communities in recent years has been in specific areas of native title, agreement making and environmental governance. Although agreement-based environmental governance forms are not unproblematic, they offer practical measures for improvement of Indigenous quality of life, especially in remote and regional areas of Australia that are more promising than in many other fields of Indigenous-non Indigenous relations (for a discussion of the role of agreements in regional and remote communities see, Gillard, 2007: 10, 13). Therefore it is useful to trace the changing dynamic of environmental governance and native title.

Native title and ecology

The interface between native title and ecology can be conceived within three key phases since the mid-twentieth century.[6] Firstly, the pre-native title era, which was characterised by the ascendancy of natural balance concepts in ecology and natural resource management, and which largely precluded Indigenous care for country apart from designated statutory schemes. This phase was accompanied by a growing momentum to involve Indigenous peoples in 'managing ecology' through cultural heritage (see, for example, *Aboriginal and Torres Strait Islander Heritage Protection Act 1984* (Cth)), statutory land rights (see, for example, *Aboriginal Land Rights (Northern Territory) Act 1976* (Cth)), and co-management schemes over national parks (see, for example, Weir, 2000) and World Heritage areas (see, for example, *Environment Protection and Biodiversity Conservation Act 1999* (Cth): Part 15 Div 4); all with various levels of responsibility accorded to Aboriginal and Torres Strait Islander peoples.

The second phase might be termed the 'Recognition and Litigation' phase of native title. Following *Mabo [No 2]* and enactment of native title legislation there was an emphasis upon defining and shaping the parameters of native title, including explorations of the 'content' of native title rights as determined by litigation involving a series of seminal High Court and Federal Court cases (for an overview of the case law, see, Strelein, 2006: Chs 1–3, 6). Given that the courts predominately favoured a view of native title as embodying non-commercial

---

6   This is a gross oversimplification as there are multilayered levels at which Indigenous peoples have a relationship with ecology as part of a wider culturally imbued understanding of place (see, for example, Rose, 1999).

interests and uses (Strelein, 2001: 95), the native title rights and interests, where such were found to be recognised and protected, largely accorded with a 'traditional' view of the relationship between Indigenous peoples and the country, the subject of native title rights. These rights implicitly included 'rights' such as care for sacred 'ecology'; particularly in regard to access to country, and to a lesser extent rights conferred in relation to its protection and management.[7]

Agreement-making as the final phase identified here, overlaps with the litigation era. Negotiated 'outcomes' for native title are becoming increasingly more prominent as the courts progressively narrowed the scope of what could be achieved for Indigenous peoples through a litigation-oriented approach, especially given the complexity, financial burden and length of litigation (Tehan, 2003: 523). Indeed, Neate suggests that agreement-making has emerged as the preferred method of dealing with native title issues (Neate, 2004: 176). Formalisation of the reliance upon agreement-making under the *Native Title Act 1993* (Cth) was established in case law: '[t]he stated emphasis of the Act [is] on the facilitation of agreement through negotiation rather then instant recourse to judicial decision' (*Fejo v Northern Territory*, 1998: Kirby J). Agreement-making of diverse levels and scope, now has assumed central importance in many spheres as defining, in legal and economic terms, how Aboriginal and Torres Strait Islander peoples' relationship with 'ecology' is managed. Indeed, agreement-making has assumed a vital role as the interface for managing many aspects of Aboriginal and Torres Strait Islander peoples 'relationship' with settler Australian society and law. Brennan and others note,

> that among the changes in the language of government is a far greater emphasis on the idea of establishing partnerships with Indigenous communities and using agreement-making as a tool of policy and administration. (Brennan *et al.*, 2005: 41)

Therefore, within the broader moves to encapsulate Indigenous peoples' relationships with country within an agreement framework, 'ecology' as it manifests in various levels will be the subject of agreement either explicitly as for example in co-management terms for identified areas of land and waters (see Szabo and Smyth, 2003), or more diffusely as part of a general stewardship accorded to native title holders, for example as access rights over a pastoral lease (see, for example, *Western Australia v Ward*, 2002). It is within this more discrete phase of agreement making this article focuses its attention.

---

7   Again the rights are highly variable on a case by case basis, ranging from 'ownership' of ecology where there is a grant of exclusive native title rights to *Native Title Act 1993* (Cth): s 211 which preserves the traditional hunting and fishing rights of Indigenous peoples but does not confer wider rights of land and water 'ownership' or exclusive possession.

Indigenous Land Use Agreements (ILUAs) under the *Native Title Act 1993* (Cth) now are integral to the native title mediation process, taking form as future act agreements,[8] or as regionally based agreements (see, for example, Aguis *et al.*, 2002), or as co-management agreements.[9] Other forms of agreements have been made between Aboriginal people and governments, non-government organisations and private entities, such as corporate businesses. Accordingly, ILUAs have assumed greater significance in the resolution of claims. Such agreements have frequently addressed land, biodiversity and cultural heritage management, particularly where part or all of the agreement creates a co-management relationship between Indigenous claimants and a parks authority (Hill, 2006: 577). Outside of the native title process, agreements have been made with Indigenous people in a growing spectrum of ecology related fields most notably under the Indigenous Protected Areas Program (see, generally, Bauman and Smyth, 2007: 13).

Thus over the last decade the proliferation of agreements between Aboriginal people, governments, non-government organisations, and private entities has seen agreement-making occupy, 'a new space between the old dichotomies of state and market, public and private, local and global' (Considine, 2005: 1). The range of agreements in place now transcends a narrow 'land use orientation' although this aspect remains important. Agreements comprise an 'emerging model of public organisation' that adopts a deregulatory middle ground between state-centred governance and privatisation, as part of policy prescriptions that have instituted 'contractualism' across many spheres of modern political social and economic life (Considine, 2005: 1). Agreement-making, is part of a broader dynamic that has emphasised the increasing pre-eminence of contractual and exchange based processes in articulating the overarching political relationship between Aboriginal and Torres Strait Islander peoples within the Australian nation state.[10] As the Federal Minister for Families, Housing, Community Services and Indigenous Affairs, Jenny Macklin observed,

> Agreements under the NTA are the major means of engagement between *Indigenous* people, industry and governments, and enable Indigenous people to plan and make decisions on a range of issues affecting their lives and their environments. (Macklin, 2009: 14–15)

---

8   *Native Title Act 1993* (Cth) Part 2 Div 3 Subdivs B, C, D and E. On the relationship between Indigenous Land Use Agreements and other 'future acts', note s 24AB(1).

9   See for example, the Wotjobaluk, Jaadwa, Jadawadjali Wergaia and Jupagulk Indigenous Land Use Agreement, signed on 11 November 2005, which forms part of the resolution of three native title claims. The Agreement, between the State of Victoria, the Commonwealth of Australia, the registered native title claimants and the Barengi Gadjin Land Council Aboriginal Corporation, provides for the grant of freehold title to three Crown allotments totalling 45 ha, National Native Title Tribunal Media Release, 25 October 2002.

10   This agreement oriented approach designated by the Howard Government as 'practical reconciliation' to date has not been substantially modified by Rudd Government policies.

Such articulation by governments of the function of agreements as 'delivering certainty' for Indigenous peoples represents a major shift in the strategic role to be played by agreement-making in providing benefits for Indigenous peoples in Australia. To date though, the major focus for securing benefits has been agreements in the mining/resource exploitation sector or the business/ entrepreneurial sphere with much less attention on benefits accruing to Indigenous peoples in the natural resources/ environment sector (Ross and Ward, 2009: 37), with the major exception of environmental co-management agreements. Co-management agreements, at least initially, were largely premised on congruence between environmental preservation and Indigenous peoples' care for country, with less attention to economic development aspects.

Agreement-making in a deregulatory environmental era

Agreement-making, native title and Indigenous environmental governance are firmly enmeshed in structural changes that are being promoted for Indigenous communities, which hinge upon economic development discourses. These discourses emphasise local Indigenous community capacity and the need to provide sustainable economic opportunities (Commonwealth of Australia, 2003). Thus agreement-making with Indigenous peoples, as a form of 'community partnership' that initially emerged in response to perceived failings of 'top down' government approaches, has now assumed a more proactive focus. Simultaneously, there have been strong moves to critique top-down statutory 'command and control' approaches in environmental governance as these models are regarded as limiting local community participation. However, despite the apparent reorientation to 'bottom up' governance in both arenas, agreement-making alone, without effective redress of the underlying structural differentials between Indigenous and non-Indigenous participants, may have limited success in establishing long term sustainable outcomes for Indigenous communities. There is a dilemma implicit to deliberative 'agreement' oriented processes where opening up dialogue between the parties can also be the mechanism to extend the policy influence of agencies involved in the agreement.

Therefore agreement-making, including ILUAs under the *Native Title Act 1993* (Cth), while proving specific opportunities for many local Indigenous communities to participate in environmental governance at a variety of levels (Bauman and Smyth, 2007) may not deal effectively with the more pervasive institutional exclusion of Indigenous peoples from economic opportunities. Indeed, historical exclusion of Indigenous peoples along the chain of policy making and service delivery has contributed much to the structural forms of disadvantage experienced by Indigenous communities (Brennan *et al.*, 2005: 35). Therefore, promotion of agreement-making in environmental governance must engage at a strategic and forward planning level (see, for discussion of a potential model, Fox, 2009: 52). Such strategies will need to negotiate the evolving 'space' between the state and the market as well as to build collaborative governance structures to achieve long-term improvements in Indigenous social capacity, environmental and economic well being. Yet when agreements are implemented in contractually-oriented policy settings, there is the potential for them to act as powerful agents of settler state economic assimilation. Accordingly, it needs to be recognised that much of the movement to embrace Indigenous partnerships for ecological protection is premised upon those communities functioning as economic as well as cultural entities under deregulatory management models (Lyster, 2002: 34, 36).

Integral to the multiple intersecting dimensions of native title, ecology and agreement-making are the economic opportunities provided by agreements in emerging areas of environmental 'service' provision (Ross and Ward, 2009: 37, 39). Environmental 'services' provision is gaining momentum as a central organising principle for many federal and state government environmental agencies; a prominent example being the introduction of the Carbon Farming Initiative by the federal government. Such economic adjustments have been clearly felt in Indigenous environmental and natural resource management spheres (Collings, 2009: 45). Altman and Cochrane endorse the conjoining of ecological protection and economic incentives by arguing for the capacity of 'bottom up' collaborative biodiversity management to achieve both economic and environmental sustainability for Indigenous communities (2005: 473). Central to this approach is a concept of 'hybridity' which integrates customary, commercial and state economic components (Altman and Cochrane, 2005: 474); also referred to as a multiplex economy (Gerritsen, 2007: 79, 81). These multiplex models are advocated for those parts of the 'Indigenous estate' which is held under some form of customary tenure or substantially controlled by Indigenous Australian – this aspect may be problematic where native title rights are held to confer less than exclusive possession (Pearson, 2004: 98, 100).[11]

Suggestions for a local hybrid economy on aboriginal land are illustrated by wildlife harvesting and management in the tropical savannah of the Northern Territory (see also, Garnett and Woinarski, 2007: 38; Gerritsen, 2007: 79), which achieves both sustainability and Indigenous economic empowerment goals. Such twining of objectives, in turn, requires new forms of interaction between Indigenous peoples, scientists and government organisations with

---

11   By contrast, Pearson has argued that, 'the common law of native title recognises that Indigenous people in occupation of land are entitled to possession where the Crown has declined to expropriate their title by act of State' (Pearson, 2004: 98, 100).

institutional innovation and purpose-built arrangements and the devolution to Indigenous community based organisations. The ramifications are far reaching, with a proposal to reorient income support for Indigenous people participating in wildlife management schemes away from, 'excessive reliance on CDEP, a work for the dole scheme' (Altman and Cochrane, 2005: 477). Similarly, '[tr]aditional indigenous skills in land and a fire management could be augmented by a role in threatened species management and exotic flora and water control to create a natural resource management economy that would be an integral part of the multiplex economy of remote Australia' (Gerritsen, 2007: 79, 81–82). With much potential for Indigenous engagement in these fields, it is necessary to canvass the factors that coalesce to produce the structural and governance changes associated with market environmentalism.

Market environmentalism

Market environmentalism has many characteristics in common with postcolonial forms of global economic policy engagement (Wallerstein, 2006: 1), particularly its relationship with an overarching discourse of economic efficiency. Under this rubric, pressures exist in many countries to introduce private property and market exchange mechanisms into customary governance of land and resources. Thus market environmentalism cannot be isolated from other political and regulatory trends that have seen the formal 'state-based' regulation of both land and Indigenous peoples (the two trends arguably are closely related) devolved to a series of intermediary and private/public partnership forms, as well as the rise of 'mutual responsibility' forms of governance associated with agreement-making.

Market environmentalism is variously linked with 'neo-liberalism', privatisation or de-regulation (see, for analysis, Beard, 2007). Broadly speaking it has precipitated a general shift in the nature of environmental governance, highlighting new discourses, such as market mechanisms and economic rationalism, which place less emphasis on the importance of a centralised role for the state in dealing with environmental problems (Stewart, 2001). This phenomenon often is described as 'governing at a distance' (Rose and Miller, 1992: 173), and is associated with the devolution of some state functions to semi-government actors and private entities who act as 'surrogate regulators' (Gunningham and Sinclair, 1998: 592, 607). In addition, it signals the expanding scope and globalisation of corporate activity into many areas of environmental protection and natural resource management (Grabosky, 1994: 419, 422). A less direct role for the state in protecting the environment and managing natural resources also can be traced to financial and resource pressures on governments (Godden and Peel, 2010: 126). These different influences have facilitated the emergence of a variety of regulatory models for environmental management in Australia which increasingly operate in conjunction with native title. However, the strategy of increasingly financially constrained governments devolving service delivery to local communities requires critical attention in the context of Indigenous participation in environmental protection and natural resource management to ensure that adequate resources are available to achieve holistic, long-term outcomes.

Such critical attention is made more pressing as economic perspectives appear to have displaced ecological science as the major contributor to understandings of how environmental problems – and indeed local communities – should be managed. Environmental problems now are conceived primarily as the problem of allocating and managing scarce resources between competing ends to achieve efficient outcomes (Ramsay and Rowe, 1995: 68). Thus in a mainstream economics framework, the environment becomes another kind of resource or asset; which provides particular types of goods and services desired by people (Tietenberg, 2004: 15). Elements of the environment become not components of inter-connected ecosystems but assets whose value is determined by utility to humans; calibrated against price. Together with biodiversity, a range of other environmental elements, such as water are now seen as providing valuable 'life-sustaining services' from fresh air and water to scenic qualities (Tietenberg, 2006: 15). Schemes in many countries have created markets for the provision of 'services' commencing with clean air and water and the avoidance of land degradation. Some schemes were developed expressly by governments[12] and others by private entities.[13] There are strong advocates of the benefits of such markets, '[t]hese experiences have demonstrated that investing in natural capital rather than built capital can make both economic and policy sense' (Salzman, 2005: 870). From an Indigenous perspective, it is important that these schemes are seen as more than investments in 'natural capital' but instead extend to long term community sustainability. Such community sustainability goals however may be eclipsed by a focus on efficiency in market environmentalism (Lyster, 2002) and an emphasis on global or national benefits to the detriment of the more localised concerns of and potential benefits for Indigenous peoples.

The public goods of environmental protection

Efficiency is measured in various ways, but broadly equates to a situation where a particular resource allocation maximises the overall benefits to society from using resources. Efficiency is given pre-eminence in economic theory, as it is

---

12   Examples of such schemes in Australia include 'Bush Tender' and 'Eco Tender' schemes in Victoria.

13   For example see Greening Australia, the organisation's website is <http://www.greeningaustralia.org.au/about-us/our-partners>

considered to be an objective criterion of social welfare (Daly and Farley, 2004: 4). Environmental economics seeks to attribute a monetary or other economic value to environmental resources or environmental protection,[14] so that the full environmental costs and benefits of resource allocations can be factored into policy and decision-making processes (Costanza *et al.*, 1998: 67), to produce a 'rational' outcome. Economic rationalism favours the free markets, with minimal governmental intervention, as the optimal means for efficient allocation of resources, in order to 'maximise' social welfare (Kinrade, 1995: 86). Yet, the belief in the free market to produce efficiency of resource allocation is underpinned by assumptions that often are problematic in terms of the public 'goods and services' provided by the environment. Situations where individual actions conflict with social objectives, are designated as market failure. Advocates of economic rationalist approaches consider that, '[t]he government's role in is to seek out market failures and correct them with policies designed to align private and social interests' (Luckert and Whitehead, 2007: 11). This constrained view of the role of government, as one limited to correcting market failures, represents a major reorientation in governance for federal and state governments in Australia. The implications of this view are already apparent in many spheres of Indigenous policy but the intersections with environmental governance are striking. If governments are seen primarily as correcting market failures, where the market itself is the main vehicle for delivery, it renders problematic many areas of strategic and institutional government service provision. Moreover, within Australia in regional and remote areas, many public goods and services are still supplied by governments, albeit with increasing levels of private provision. In many instances though, private provision will be uneconomic and/or unable to address externalities and third party effects. Externalities such as pollution and third party effects, that is effects on entities that are not parties to the market exchange/agreement, are common in environmental spheres. Typically, much private exchange based environmental service provision does not address these aspects – these are the so-called market failures.[15] To withdraw government environmental protection and broader service provision where market failures operate thus risks increasing Indigenous disadvantage and may lock regions into a 'resource curse' situation where the benefits of natural resource/economic exploitation flow outside the source region.[16] Accordingly, it is necessary to consider the circumstances where governments should adopt a more expansive role in promoting economic development and stimulus consistent with broader

---

14   Adam Smith's *The Wealth of Nations* was a seminal work advocating market mechanisms to achieve efficiency in allocation.

15   As a consequence, private entities often are able to 'cherry pick' the more profitable aspects of environmental service provision.

16   For an analysis of the resource curse concept see Langton and Mazel, 2008: 31–33. While this phenomenon is most often associated with mineral exploitation, it is apparent that NRM can be subject to similar disparities in return income flow.

social goals, such as the alleviation of the relative disadvantage of Indigenous communities. This expansive role is most imperative in areas of 'public goods' and ecosystem services such as biodiversity protection.

Markets tend to perform poorly when it comes to the allocation of environmental resources that are public goods (Moran, 1995: 73, 79), or so called 'common pool resources' – including many areas that comprise Indigenous land and waters. NRM is often characterised by market failures for public goods, as the 'goods' obtained such as climate mitigation or land protection cannot be regarded as 'exclusive'; that is third parties cannot be effectively excluded from the benefits of such environmentally beneficial measures, even though such entities may not contribute to the provision of such benefits (the free rider problem). Luckert and Whitehead suggest,

> there are many goods and services that suffer from public good properties. In addition to biodiversity, carbon sequestration, retention of cultural values and scenic resources may all exhibit public group properties. Many of these resources are enjoyed by non-indigenous Australians who are able to free ride on the provision of these values by indigenous peoples. (Luckert and Whitehead, 2007)

Given that many native title determinations do not confer exclusive possession on native title holders – often such rights will comprise a non-exclusive right of access and use – then the free rider problem may be exacerbated in these circumstances. The situation may be less critical under statutory land rights schemes and in 'joint management' situations where there is greater Indigenous control over access. However, incursions into Indigenous control over access to land and resources, such as that under the Northern Territory 'intervention'[17] must be treated with caution where there are attempts to facilitate 'economic opportunities' without appropriate safeguards for Indigenous management of lands and resources. By contrast, market failure and problems in the exclusion of free riders offer a grounded economic rationale justifying government programs promoting Indigenous involvement in NRM and environmental protection programs (Luckert and Whitehead, 2007).[18] The non-government sector also can be involved in such programs (O'Riordan, 1981: ix), – again an important consideration for future directions in ecology, native title and partnership approaches. Such co-regulatory approaches where governments promote market

---

17 Pursuant to s122 of the *Australian Constitution* ('Territories Power') the Commonwealth 'intervened' to address perceived child welfare and human rights abuses in the Northern Territory. See *Northern Territory National Emergency Response Act 2007* (Cth), *Social Security and Other Legislation Amendment (Welfare Payment Reform) Act 2007* (Cth); *Families, Community Services and Indigenous Affairs and other Legislation Amendment (Northern Territory National Emergency Response and other Measures) Act 2007* (Cth).

18 These authors suggest that there are two principles on which such programs should be predicated; first an 'impacter pays' principle and secondly, a 'beneficiary pays' principle.

opportunities in environmental management and biodiversity protection are welcome additions to the range of environmental regulation where they give due acknowledgment to Indigenous autonomy and long term community viability.

Trends are less encouraging in relation to the recognition of Indigenous peoples and local community governance of 'common pool resources', at an international level. In the case of the common pool resources, classical economic theories, predicated upon Coase's social cost theorem (Coase, 1960: 1), argue that 'informal users' deplete the 'common resource' without regard to the interests of the broader community. Ultimately this situation will result in over-exploitation and collapse (Hardin, 1968: 1243).[19] Informal users are regarded as those people not part of a formal private property regime. Solutions that are advocated to deal with this type of 'market failure' generally suggest making land and environmental resources a private entity and tradeable commodity. A prominent recent example of dealing with the degradation of common pool resources in this manner is through the creation of emission permits in an emissions trading market and in carbon 'offsets' (see, for analysis, Gerrard, Chapter 7, this volume). Drawing on social cost theory and the proliferation of cap and trade schemes, market mechanisms for trading in environmental resources, such as water, have been adopted in environmental regulation in many parts of Australia.

The use of market tools for the purposes of environmental regulation has not occurred without substantial critiques. Yet market environmentalism continues to gain momentum as 'market tools' have been strongly promoted by a suite of new (and old!) governmental institutions, increasingly committed to the property rights/trading model.[20] Indeed, use of trade and exchange to regulate for environmental goals is regarded now in policy and government circles as offering substantial advantages compared with traditional state based ecologically oriented environmental regimes (see, for discussion, Ackerman and Stewart, 1985: 1333). However as experience with market environmentalism emerges, the Australian situation suggests that markets in practice do not always capture the predicted theoretical benefits (Eckersley, 1995: 7, 21). Generally, what is emerging in Australia as the predominant market-based model is not 'free' market, but rather a hybrid of command-and-control and market measures perhaps best categorised as 'legally regulated marketization' (Braithwaite and Parker, 2004: 269). Specific criticisms of the market model of particular pertinence to this analysis are that market exchange may work effectively for environmental regulation in situations of discrete pollutants which can be

---

19 Hardin's article generated many counter views that challenge the premise that human beings inevitably act as individualised rational economic actors incapable of organising communitarian responses to protect shared environmental resources (see, for example, Ostrom, 1990).

20 For example, the Productivity Commission has issued many reports of the potential for markets in 'ecosystems services' in areas such as water management, salinity control, biodiversity conservation and carbon capture (see, for example, Murtough *et al.*, 2002; Productivity Commission, 2001).

effectively monitored and valued for trade/offset purposes, (although even here not all costs are captured). However trade and exchange are less effective for complex, heterogeneous environmental resources such as biodiversity (Luckert and Whitehead, 2007: 11). Other critiques of the market-based model point to more intangible values that may be sacrificed through the focus of markets on the criterion of efficiency (Jacobs, 1995: 46, 68). These points have particular resonances for the intersections between native title, ecology and markets given that many western legal and economic instruments fail to capture the nuances of Indigenous relationships with country.

Markets, ecology and native title

How then are Indigenous peoples, native title, ecology to be situated in the rapidly emerging regulatory structures of environmental markets and the accompanying structural changes? In turn, how should agreement-making operate in such a governance and organisational space?

Clearly, significant potential exists to utilise native title and agreement-making in concert with sustainable environmental objectives to facilitate long term structural change in the economic opportunities available to Indigenous communities (see, for example, Ridgeway, 2005), particularly, but not exclusively, in remote and regional Australia. Thus,

> [t]he insertion of customary institutions and jurisdictions into the market place through agreement making, such as Aboriginal heritage management agreements ... is not mere syncretisation of tradition and modernity, but the transformation of relationships. These postcolonial forms of policy engagement are underwritten by both customary exchange and market considerations. (Langton and Palmer, 2004: 47)

Postcolonial policy engagements combining market considerations and forms of exchange built upon Indigenous knowledge and values governing care for country are exemplified, as noted, by Indigenous agreements for the provision of a variety of ecosystems services, such as carbon sequestration. (Gerrard, 2008: 941, 945). As Gerrard suggests:

> Many environmental services performed by Indigenous peoples are not 'new' to federal, State and Territory governments. Government departments and agencies have been involved in joint and cooperative management arrangements with Indigenous peoples for some time. However, the current threat of climate change and associated 'low carbon'

context creates the need to value these services more appropriately and to provide adequate financial and regulatory infrastructure to enable access to, and growth of, new opportunities. (Gerrard, 2008: 945)

Indigenous involvement in the provision of environmental and NRM services reflects structural and administrative governance changes that have direct consequences for Indigenous employment and benefit flow to Indigenous communities. A further advantage to stem from such involvement is the potential for closer integration of Australian Indigenous communities with global markets in environmental 'goods'. The predicted rise of green economies in a carbon constrained world arguably presents unique opportunities for Indigenous peoples whose customary knowledge of land and waters can make a significant contribution to international and national efforts to address environmental deterioration including climate change. Yet, as Gerrard comments in regard to climate change mitigation and adaptation;

> [a]lthough parts of Australia have benefited from innovative and supportive Caring for Country programs and Indigenous environmental and land management services, greater acknowledgement and support is needed for Indigenous peoples to grow development opportunities associated with climate mitigation activities. At present, traditional knowledge and the ecological services performed by Indigenous peoples are generally informal, undervalued and/or under-supported. (Gerrard, 2008: 941)

Indigenous peoples' involvement in environmental services must be predicated upon an approach that respects and gives effect to Indigenous people's customary relationship with land and waters but which is contemporary and practically grounded (Ross and Ward, 2009: 39). Experience with agreement-making in Australia and comparable jurisdictions suggests that effectively negotiated and implemented agreements can offer a structure to value Indigenous contributions to environmental management both tangible and intangible and facilitate access to, and growth of new opportunities for Indigenous communities (see, generally, Langton *et al.*, 2006). Environmental markets provide Indigenous communities with the chance to participate in rural businesses which are uniquely located in regional and remote areas. The lack of inherently viable local business in such areas has long been acknowledged as problematic on the basis that 'traditional' forms of capital investment are not available.

However, agreement-making in environmental markets must also operate within the ambit of strong national legislative and international law safeguards for traditional knowledge and customary practice in the delivery of environmental services.[21] The history of the interface between Indigenous communities, traditional knowledge and values, and global markets has not always been beneficial to Indigenous peoples (see, for example, Davis, 1999: 40).[22] Global markets in biotechnology are now pervasive, supplying many goods that have a basis in either genetic material gained from Indigenous held land and waters or utilising traditional knowledge. Given this experience, clearly there is a need to develop robust protocols and legally enforceable safeguards to protect Indigenous peoples and traditional knowledge practices that are integral to Indigenous participation in 'environmental services' provision.

From the perspective of changing structural forms and employment opportunities, earlier experience of globalisation and the market penetration of Indigenous communal systems, gives some cause for concern. The adoption of market environmentalism offers some parallels with earlier periods of globalisation and privatisation that intruded into Indigenous capacity to care for country (Gerritsen, 2007: 79). Gerritsen argues that previous interventions of western capitalist modes of production into remote and regional parts of Australia also were predicated upon the idea of the identified need for greater efficiencies. However a rubric of efficiency often acted to displace Indigenous peoples from the labour force and emerging economies (Gerritsen, 2007: 79). Alternatively, in market environmentalism, are Indigenous peoples likely to become a 'labour force' in yet another industry where Indigenous employment is concentrated at the unskilled 'end', rather than at the managerial/decision making level? The history of the pastoral industry offers salutary historical experience in this regard (Gerritsen, 2007: 79). In this context, market environmentalism might be regarded as yet another capitalist encroachment; a type of 'green-wash' intervention that needs to be carefully managed to ensure that full participation of Indigenous communities and a flow of benefits to communities from the use of Indigenous land and waters and associated traditional knowledge is achieved. As noted, careful monitoring is important especially where financially constrained governments may be tempted to regard Indigenous participation in ecosystem services as a lower-cost option. In the context of global warming Gerrard this volume identifies that,

> Australia's responses to climate change must preserve space for Indigenous peoples to determine and realise meaningful opportunities based on their specialised knowledge and traditional practices. (Gerrard, Chapter 7, this volume)

---

21   The important role of traditional knowledge in biodiversity conservation is recognised explicitly in the *Convention on Biological Diversity* articles 8(j), 10(c), 17(2) and 18(4).

22   Indigenous peoples' experience with global biotechnology markets and the exploitation of traditional knowledge being a case in point.

Further, it may be useful to consider the other forms of institutional and structural reforms, such as taxation incentives, that may be required to fully implement a robust model for effective Indigenous employment and business development under the auspices of agreement-making for environmental service delivery.

Conclusion

In many countries where indigenous peoples and local communities hold significant amounts of communal land and resources, these customary systems increasingly are regarded as problematic and 'inefficient' in persuasive neo-liberal policy platforms. There are strong pressures operating through globalisation forces, and in equivalent domestic national natural resource and environmental management policies to renounce communal holding in favour of market and property-based regimes (Hughes and Warin, 2005). When aligned with an uncritical acceptance of the 'Tragedy of the Commons' phenomenon, it produces the call to formalise and individualise communal governance of land and waters, often irrespective of the local situations that pertain (Schlager and Ostrom, 1993: 14–18; see, more generally, Ostrom, 1990). Importantly, adoption of 'the commons' terminology also blurs together many facets of multilayered Indigenous governance systems under a simplifying assumption that imports a bimodal schema of either private or communal categories (see, for a critique, Lee, 2006: 22). Such oppositional categories fail to accommodate the highly porous nature of relationships that exist in Indigenous and local communities which allows diverse forms of entitlement and responsibility within an overarching communal governance system (Macintyre and Foale, 2007: 49–59).

On an international scale, communal governance of land and resources that, broadly speaking, can be regarded as equivalent to native title, is subject to internal policy and externally derived trends to replace such tenures with commercially oriented, market-based forms as the preferred medium for Indigenous peoples to manage land and resources. Under this trend, complex interactions with land and waters, that offer the capacity for more multifaceted objectives in protecting ecological and spiritual relationships, risk being constituted as inefficient. Inefficiency is the most recent signification of a long history of property relations between Indigenous peoples and settler societies that have inscribed Indigenous relationships with land as ineffective or even more negatively as non existent (that is as terra nullius). In concert, long standing discourses of security and certainty have produced a sense that the modern, highly technological use of land – that which creates settler property – is commercial and private. Assigning causal trajectories, such as inefficient land and resource practices with customary systems, reveals the limitations of market values and efficiency paradigms for holistic ecological outcomes, even where these have been substantially modified to take account of cultural factors. Moreover, what is at stake in many debates over market regulation is not only economic productivity, but also the struggle for recognition of more dynamic community relationships than those modelled on market environmentalism. Finally, a key issue is whether the instigation of agreement-making in environmental markets deflects from fundamental questions about Indigenous self determination and native title as denoting ownership of land and communal resources? More positively perhaps, the challenges and opportunities presented by the change in paradigm from ecology to environmental markets are ones to be grasped by Indigenous communities. As the policy and legal responses to climate change and ecological preservation that are emerging around carbon sequestration and fire burning practices signal, there are significant windows of opportunity opening up for Indigenous Australians to participate in major new structural models for ecological management and benefit. Such change is precipitating under a range of ecological imperatives such as climate change adaptation, but also due to significant shifts in international finance and trading regimes, and indeed to the very concepts of value and 'offset' in a range of ecologically-related activities. As these emerging models of governance over land and resources crystallise, institutional, legal, political and cultural questions will arise as to how to most effectively provide a platform for recognising the significance of Indigenous relationships with country and the need for flexibility and local community participation in native title and associated forms of agreement-making with Indigenous peoples.

References

Ackerman, BA and RB Stewart 1985, 'Reforming environmental law', *Stanford Law Review* 37(5): 1333–1365.

Aguis, P, J Davies, R Howitt and L Johns 2002, '2004, 'Negotiating comprehensive settlement of native title issue: building a new scale of justice in South Australia', *Land Rights and Native Title* 2(20): 1–12.

Altman, J and M Cochrane 2005, 'Sustainable development in the Indigenous-owned savanna: innovative institutional design for cooperative wildlife management', *Wildlife Research* 32(5): 473–480.

Anderson, T and R Simmons (eds) 1993, *The Political Economy of Customs and Culture: Informal Solutions to the Commons Problem*, Rowman and Littlefield, Lanham.

Bauman, T and D Smyth 2007, *Indigenous Partnerships in Protected Area Management: Three Case Studies*, Australian Institute of Aboriginal and Torres Strait Islander Studies, Canberra.

Beard, J 2007, *The Political Economy of Desire: International Law, Development and the Nation State*, Routledge-Cavendish, Oxon.

Blaser, M, HA Feit and G McRae 2004, 'Indigenous peoples and development processes: new terrains of struggle', in *In the Way of Development: Indigenous Peoples, Life Projects and Globalization*, M Blaser, H Feit and G McRae (eds), Zed Books, London: 1–25.

Braithwaite, J and C Parker 2004, 'Conclusion', in *Regulating Law*, C Parker, C Scott, N Lacey and J Braithwaite (eds), Oxford University Press, Oxford: 269–289.

Brennan, S, L Behrendt, L Strelein and G Williams 2005, *Treaty*, The Federation Press, Sydney.

Coase, R 1960, 'The problem of social cost', *Journal of Land and Economics* 3(1): 1–44.

Collings, N 2009, 'Native title, economic development and the environment', *Reform* 93: 45–47.

Commonwealth of Australia 2003, *Shared Responsibility Through Partnership: Indigenous Issues Fact Sheet Series*, available at: <http://www.Indigenous.gov.au/ip.dll/SearchResults?s=1427> (accessed 8 November 2004).

Considine, M 2005, *Partnerships and Collaborative Advantage: Some Reflections on New Forms of Network Governance*, Centre for Public Policy, The University of Melbourne, Melbourne.

Costanza, R, R d'Arge, R de Groot, S Farber, M Grasso, B Hannon, K Limburg, S Naeem, R O'Neill, J Paruelo, R Raskin, P Sutton and M van den Belt 1998, 'The value of ecosystem services: putting the issues in perspective', *Ecological Economics* 25(1): 67–72.

Daly, HE and J Farley 2004, *Ecological Economics: Principles and Applications*, Island Press, Washington DC.

Davis, M 1999, 'Indigenous rights in traditional knowledge and biological diversity: approaches to protection', *Australian Indigenous Law Reporter* 4(4): 1–32.

Department of Environment and Heritage (DEH) 2001, 'Indigenous people and biodiversity', *State of the Environment Report 2001*, Department of Environment and Heritage, Canberra, available from <http://www.environment.gov.au/soe/2001/index.html> (accessed 5 October 2010).

DEH – *see* Department of Environment and Heritage

DEWHA – *see* Department of Environment, Water, Heritage and the Arts

Dovers, S and R Price 2007, 'Research and the integration imperative', in *Integrated Resource and Environmental Management: Concepts and Practice*, KS Hanna and DS Slocombe (eds), Oxford University Press, Oxford and Toronto: 36–55.

Ecologically Sustainable Development Steering Committee 1992, *National Strategy for Ecologically Sustainable Development*, prepared by the Department of Environment, Water, Heritage and the Arts (DEWHA), Canberra.

Eckersley, R 1995, 'Markets, the state and the environment: an overview', in *Markets, the State, and the Environment: Towards Integration,* R Eckersley (ed), Macmillan, Melbourne: 7–45.

Fisher, D 1999, 'The impact of international law upon the Australian environmental legal system', *Environmental & Planning Law Journal* 16(5): 372–374.

Fox, P 2009, 'Indigenous land use agreements: a Canadian model', *Reform* 93: 52–54.

Garnett, ST and CZ Woinarski 2007, 'A case for Indigenous threatened species management', in *Investing in Indigenous Natural Resource Management,* MK Luckert, BM Campbell, JT Gorman and ST Garnett (eds), Charles Darwin University Press, Darwin: 227–259.

Gerrard, E 2008, 'Climate change and human rights: issues and opportunities for indigenous peoples', *University of New South Wales Law Journal* 31(3): 941–952.

Gerritsen, R 2007, 'A case for Indigenous employment in NRM', in *Investing in Indigenous Natural Resource Management,* MK Luckert, BM Campbell, JT Gorman and ST Garnett (eds), Charles Darwin University Press, Darwin: 79–85.

Gillard, J 2007 'Time to end the culture war', *Australian Quarterly* 79(4): 10–13.

Godden, L and J Peel 2010, *Environmental Law: Scientific Policy and Regulatory Perspectives,* Oxford University Press, Melbourne.

Grabosky, P 1994, 'Green markets: environmental regulation by the private sector', *Law & Policy* 16(4): 419–448.

Gunningham, N and D Sinclair 1998, 'New generation environmental policy: environmental management systems and regulatory reform', *Melbourne University Law Review* 22(3): 592–616.

Hardin, G 1968, 'The tragedy of the commons', *Science* 162(3859): 1243–1248.

Hill, R 2006, 'The effectiveness of agreements and protocols to bridge between Indigenous and Non Indigenous toolboxes for protected area management: a case study from the Wet Tropics of Queensland', *Society and Natural Resources* 19(7): 577–590.

— and L Williams 2009, 'Indigenous natural resource management: overcoming marginalisation produced in Australia's current NRM model', in *Contested Country: Local and Regional Natural Resources Management in Australia*, M Lane, C Robinson and B Taylor (eds), CSIRO Publishing, Collingwood: 161–178.

Hokari, M 2005, 'Gurinndji mode of historical practice', in *The Power of Knowledge: The Resonance of Tradition*, L Taylor, G Ward, G Henderson, R Davis and L Wallis (eds), Aboriginal Studies Press, Canberra: 214–222.

Hughes, H and J Warin 2005, *A New Deal for Aborigines and Torres Strait Islanders in Remote Communities,* Centre for Independent Studies, St Leonards, New South Wales.

Jacobs, M 1995, 'Sustainability and "the market": a typology of environmental economics', in *Markets, the State, and the Environment: Towards Integration*, R Eckersley (ed), Macmillan, Melbourne: 46–71.

Kinnane, S 2005, 'Indigenous sustainability: rights, obligations and a collective commitment to country', in *International Law and Indigenous Peoples*, J Castellino and N Walsh (eds), Kluwer Law International, Leiden and Boston: 195–224.

Kinrade, P 1995, 'Towards ecologically sustainable development: the role and shortcomings of markets', in *Markets, the State, and the Environment: Towards Integration,* R Eckersley (ed), Macmillan, Melbourne: 86–111.

Langton, M 2003, 'The "wild", the market and the native: indigenous people face new forms of global colonization', in *Globalization, Globalism, Environments, and Environmentalism: Consciousness of Connections (The Linacre Lectures)*, S Vertovec and D Posey (eds), Oxford University Press, Oxford: 141–169.

— and O Mazel 2008, 'Poverty in the midst of plenty: Aboriginal people, the resource curse and Australia's mining boom', *Journal of Energy and Natural Resources Law* 26(1): 31–65.

—, O Mazel, L Palmer, K Shain and M Teehan (eds), 2006, *Settling with Indigenous People: Modern Treaty and Agreement-making,* Federation Press, Sydney.

— and L Palmer 2004, 'Treaties, agreement making and the recognition of indigenous customary polities', in *Honour Among Nations: Treaties and Agreements with Indigenous People,* M Langton, L Palmer, M Tehan and K Shain (eds), Melbourne University Press, Carlton: 34–50.

—, ZM Rhea and L Palmer 2005, 'Community-oriented protected areas for indigenous peoples and local communities', *Journal of Political Ecology* 12: 23–50.

Lee, P 2006, 'Individual titling of Aboriginal land in the Northern Territory: what Australia can learn from the international community', *University of New South Wales Law Journal* 29(2): 22–38.

Luckert, MK and PJ Whitehead 2007, 'A general case for natural resources management: market failures and government policy', in *Investing in Indigenous Natural Resource Management,* MK Luckert, B Campbell, JT Gorman and S Garnett (eds), Charles Darwin University Press, Darwin: 11–18.

Lyster, R 2002, '(De)regulating the rural environment', *Environmental and Planning Law Journal* 19(1): 34–57.

Macklin, J 2009, 'Can native title deliver more than a modicum of justice?', *Reform* 93: 14-16.

Macintyre, M and S Foale 2007, 'Land and marine tenure, ownership and new forms of entitlement on Lihir: changing notions of property in the context of a gold mining project', *Human Organisation* 66(1): 49–59.

Massey, D 2005, *For Space,* Sage, London.

Moran, A 1995, 'Tools of environmental policy: market instruments versus command-and-control', in *Markets, the State, and the Environment: Towards Integration,* R Eckersley (ed), Macmillan, Melbourne: 73–85.

Murtough, G, B Aretino and A Matysek 2002, *Creating Markets for Ecosystem Services,* Productivity Commission Staff Research Paper, AusInfo, Canberra.

Neate, G 2004, 'Agreement making and the Native Title Act', in *Honour Among Nations: Treaties and Agreements with Indigenous People,* M Langton, L Palmer, M Tehan and K Shain (eds), Melbourne University Press, Carlton: 176–189.

Nettheim, G, GD Meyers and D Craig 2002, *Indigenous Peoples and Governance Structures: A Comparative Analysis of Land and Resource Management Rights,* Aboriginal Studies Press, Canberra.

O'Riordan, T 1981, *Environmentalism*, 2nd edition, Pion, London.

Ostrom, E 1990, *Governing the Commons: The Evolution of Institutions for Collective Action*, University of Cambridge Press, Cambridge.

Pearson, N 2004, 'Land is susceptible of ownership', in *Honour Among Nations: Treaties and Agreements with Indigenous People,* M Langton, L Palmer, M Tehan and K Shain (eds), Melbourne University Press, Carlton: 83–101.

Peel, J 2008, 'Ecologically sustainable development: more than mere lip service?', *Australasian Journal of Natural Resources Law & Policy* 12(1): 1–34.

— and L Godden 2005, 'Australian environmental management: a dams story', *University of New South Wales Law Journal* 28(3): 668–695.

Plumwood, V 2003, 'Decolonizing relationships with nature', in *Decolonizing Nature: Strategies for Conservation in a Post-colonial Era,* WM Adams and M Mulligan (eds), Earthscan, London: 51–78.

Productivity Commission, 2001, 'Harnessing Private Sector Conservation of Biodiversity: Commission Research Paper' AusInfo, Canberra, available at: <http://www.pc.gov.au/?a=8203>

Ramsay, R and GC Rowe 1995, *Environmental Law and Policy in Australia: Text and Materials*, Butterworths, Sydney.

Ridgeway, A 2005, 'Addressing the economic exclusion of Indigenous Australians through native title', The Mabo Lecture, Paper presented at the National Native Title Conference, Coffs Harbour, April 2005.

Rose, N and P Miller 1992, 'Political power beyond the state: problematics of government', *British Journal of Sociology* 43(2): 173–205.

Rose, DB 1999, 'Indigenous ecologies and an ethic of connection', in *Global Ethics and Environment,* N Low (ed), Routledge, London: 175–187.

Ross, S and N Ward 2009, 'Mapping Indigenous peoples' contemporary relationships to country', *Reform* 93: 37-40.

Salzman, J 2005, 'Creating markets for ecosystem services: notes from the field', *New York University Law Review* 80: 870–961.

Schlager, E and E Ostrom 1993, 'Property-rights regimes and coastal fisheries: an empirical analysis', in *The Political Economy of Customs and Culture; Informal Solutions to the Commons Problem,* T Anderson and R Simmons (eds), Rowman and Littlefield, Manham, MD: 13–42.

Stewart, RB 2001, 'A new generation of environmental regulation?', *Capital University Law Review* 29(21): 21–182.

Strelein, L 1993, 'Indigenous peoples and protected landscapes in Western Australia', *Environmental & Planning Law Journal* 10(6): 380–397.

— 2001, 'Conceptualising native title', *Sydney Law Review* 23: 95–124.

— 2004, 'Symbolism and function: from native title to Aboriginal and Torres Strait Islander self-government', in *Honour Among Nations: Treaties and Agreements with Indigenous People,* M Langton, L Palmer, M Tehan and K Shain (eds), Melbourne University Press, Carlton: 189–203.

— 2006, *Compromised Jurisprudence: Native Title Cases since Mabo,* Aboriginal Studies Press, Canberra.

Szabo, S and D Smyth 2003, 'Indigenous Protected Areas in Australia', in *Innovative Governance: Indigenous Peoples, Local Communities and Protected Areas,* H Jaireth, and D Smyth (eds), Ane Books, New Delhi: 145–164.

Tehan, M 2003, 'A hope disillusioned, an opportunity lost? Reflections on common law native title and ten years of the Native Title Act', *Melbourne University Law Review* 27(2): 523–570.

—, L Palmer, M Langton and O Mazel 2006, 'Sharing land and resources: modern agreements and treaties with indigenous peoples in settler states', in *Settling with Indigenous People: Modern Treaty and Agreement-making,* M Langton, O Mazel, L Palmer, K Shain and M Tehan (eds), Federation Press, Sydney: 1–18.

Tietenberg, T 2004, *Environmental Economics and Policy,* Addison Wesley, Boston.

— 2006, *Environmental and Natural Resource Economics,* Addison Wesley, Boston.

Thorley, P 2002, 'Current realities, idealised pasts: archaeology, values and Indigenous heritage management in Central Australia', *Oceania* 73(22): 110–126.

United Nations Environment Program 1994, *Convention on Biological Diversity: Text and Annexes*, UNEP/CBD/94/1, Geneva: Interim Secret Am. Biol. Divers, United Nations, New York.

Verran, H 2008, 'Science and the Dreaming', *Issues* 82: 23–27.

Wallerstein, I 2006, *European Universalism: The Rhetoric of Power,* New Press, New York.

Weir, J 2000, 'Aboriginal land ownership and joint management of national parks in NSW', *Indigenous Law Bulletin* 5(3): 20–22.

West, P, J Igoe and D Brockington 2006, 'Parks and peoples: the social impact of protected areas', *Annual Review of Anthropology* 35: 251–277.

Worster, D 1994, *Nature's Economy: A History of Ecological Ideas,* University of Cambridge Press, Cambridge.

Archival source

Agreements, Treaties and Negotiated Settlements Database, University of Melbourne, available at: <http://www.atns.net.au/> (accessed 9 May 2009).

Cases

Fejo v Northern Territory (1998) 195 CLR 96, 139, Kirby J

Mabo v Queensland [No 2] (1992) 175 CLR 1

Members of the Yorta Yorta Aboriginal Community v Victoria (2002) 214 CLR 422

R v Murphy (1990) 71 LGRA 1

Western Australia v Ward (2002) 213 CLR 1, [49]

Yanner v Eaton (1999) 201 CLR 351

Legislation

Aboriginal and Torres Strait Islander Heritage Protection Act 1984 (Cth)

Aboriginal Land Rights (Northern Territory) Act 1976 (Cth)

Environment Protection and Biodiversity Conservation Act 1999 (Cth) Part 15 Div 4

Families, Community Services and Indigenous Affairs and other Legislation Amendment (Northern Territory National Emergency Response and other Measures) Act 2007 (Cth)

Native Title Act 1993 (Cth)

Northern Territory National Emergency Response Act 2007 (Cth)

Social Security and Other Legislation Amendment (Welfare Payment Reform) Act 2007 (Cth)

7. Towards a Carbon Constrained Future: Climate Change, Emissions Trading and Indigenous Peoples' Rights in Australia[1]

Emily Gerrard

Our traditional knowledge on sustainable use, conservation, protection of our territories has allowed us to maintain our ecosystems in equilibrium ... Our cultures, and the territories under our stewardship, are now the last ecological mechanisms remaining in the struggle against climate devastation. All Peoples of the Earth truly owe a debt to Indigenous Peoples for the beneficial role our traditional subsistence economies play in the maintenance of planet's ecology. (Declaration of Indigenous Peoples on Climate Change, 2000: Articles 2 and 3)

Despite the turbulent evolution of climate change law and policy over recent years, the opportunities, issues and risks for Indigenous peoples arising from the use of market based mechanisms to address environmental issues remain relatively unchanged. The central theme of this paper (first drafted in 2008) also remains unchanged: the importance of early and meaningful engagement with, and respect for, Indigenous peoples and their rights, interests and knowledge in this rapidly evolving area of law and policy.

Since the initial version of this paper in 2008, the Carbon Pollution Reduction Scheme has been proposed and defeated and the political landscape has shifted to a minority Labor Government. There have been a number of climate change conferences and developments in greenhouse gas regulation, both nationally and internationally. This includes the recent *Carbon Credits (Carbon Farming Initiative) Act 2011* (Cth), which passed through the Australian Parliament on 23 August 2011.

Key updates have been made to this paper to reflect the contemporary policy announcements and proposed regulatory instruments within Australia.

---

1 This paper was originally written subsequent to a presentation on climate change at the AIATSIS Native Title Conference in June 2007 (in part published as Gerrard, 2008). The reference list includes a number of works that are not cited in the text but form part of the background research for the original paper and the current work and are considered a useful resource for the reader.

However, at the time of writing this prologue, Australia sits on the cusp of further significant reform: the Clean Energy Future Plan (and carbon pricing mechanism – or carbon tax) of the Gillard Labor Government.

Developments at the international level include a number of non-binding commitments by parties to the United Nations Framework Convention on Climate Change (UNFCCC) and Kyoto Protocol at Copenhagen and Cancun. In the absence of a binding international agreement, the obligations of various international countries remains uncertain beyond the Kyoto Protocol's first commitment period (2008–2012) – the 'post 2012' environment.

Climate change has focused global attention on the multiple important values of our natural environment and fragile ecosystems. Within this shifting focus lie opportunities for Indigenous people, whose knowledge, understandings and practices, not to mention landholdings, hold great potential for solutions to the problems facing our natural world.

Debate about climate change has been intensifying at an international level for decades. In Australia, momentum has grown rapidly since 2007 with ratification of the Kyoto Protocol, the Garnaut Reviews in 2008 and 2011 (Garnaut, 2008, 2011),[2] and the proposed introduction of regulation and trading mechanisms to address greenhouse gases emissions (DCC, 2008a, 2008b, 2009b; DCCEE 2011a, 2011b).[3] However, while governments have improved their approach since 2007, these responses have not adequately engaged with Indigenous Australians.

Maximising space for Indigenous participation and economic development in emerging law and policy relies on the recognition of fundamental rights and interests of Indigenous peoples. This includes the right to participate in development opportunities in accordance with Indigenous peoples' needs, interests and aspirations. It also includes the right to determine and realise meaningful opportunities based on their specialised knowledge and traditional practices. Indigenous peoples' rights and interests must be embedded in emerging climate change law and policy in Australia to avoid another frontier for assimilation and appropriation of country, knowledge and culture.

---

2   Action since 2007 was preceded by coordinated State and Territory government responses, for example the National Emissions Trading Taskforce, established by Australian State and Territory Governments in 2004 in response to the Australian Government's refusal to ratify the Kyoto Protocol. The Taskforce was formed to examine proposed design options for a multi-jurisdictional Australian emissions trading scheme. The final report of the Taskforce in December 2007 was released for consideration by the Garnaut Climate Change Review process, see: <http://www.garnautreview.org.au/CA25734E0016A131/WebObj/ NETTReportfromStateandTerritoryOfficials_Finalreceived14March2008/$File/NETT%20Report%20 from%20State%20and%20Territory%20Officials_Final%20received%2014%20March%202008.pdf>
3   Australian Government Carbon Pollution Reduction Scheme Green Paper/White Paper consultation process (July – December 2008) and exposure draft legislation (March 2009), CFI consultation and subsequent legislation, December 2010 and the recent Clean Energy Future 2011 package of reform. See: <http:// www.climatechange.gov.au/cfi> (accessed 12 August 2011), and <http://www.cleanenergyfuture.gov.au/> (accessed 12 August 2011).

This paper explores the potential opportunities for Indigenous peoples in the emerging carbon economy and issues at the interface between western concepts of property and law and the understandings, values and cultural responsibilities of Indigenous people. In exploring these issues, it is acknowledged that participating in the carbon economy may not be considered appropriate by all Indigenous peoples.

Climate change – evolving global awareness

In 1979, an international conference convened by the World Meteorological Organisation expressed concern that human activities were causing regional and perhaps global changes in climate. The conference called for global cooperation to examine the future direction of climate change and appealed to nations to foresee and prevent changes in climate that may be adverse to the well-being of humanity (IPCC, 2004).

In 1985 a joint World Meteorological Organisation, United Nations Environment Programme and International Council for Science concluded that global average temperatures were likely to rise by the first half of the 21st century as a result of increased greenhouse gases from human activities (IPCC, 2004).

In 1988 the World Meteorological Organisation and United Nations Environment Programme established the Intergovernmental Panel on Climate Change (IPCC) to examine, monitor and report on matters relating to climate change. Early work of the IPCC underpinned the drafting and signing of the UNFCCC in 1992.

The UNFCCC commenced in 1994 and provides a mechanism for intergovernmental action to address climate change. As parties to the UNFCCC, governments agree to gather and share information on greenhouse gas emissions, policy responses and best practices. Parties also agree to introduce strategies for addressing greenhouse gas emissions and adapting to the expected impacts of climate change (UNFCCC, 1994).

Scientific certainty in relation to global warming and the role of human activities in accelerating climate change has increased since early reports of the IPCC (2007).[4] In February 2007 the IPCC Working Group 1 reported that global temperatures may rise from 1.1 to 6.0° Celsius by 2100, and sea levels from 18 to 59 centimetres, depending on future greenhouse gas emissions (IPCC, 2007). Scientists and governments have since warned that sea level rise is likely to surpass these estimates and reach possible levels of 140 centimetres

---

4   The IPCC concluded in 2007 that there was a 'very high confidence' (greater than 90 per cent chance) that global warming is occurring as a result of human activities.

(DCC, 2008b: 2–3).[5] More recently, Professor Garnaut has noted that advances in science about the warming of the earth and the contribution of human activity to this phenomenon as beyond reasonable doubt (Garnaut, 2011).

In Australia warming is likely to occur at a rate similar to average global temperature increases. As a result of reduced rainfall and increased evaporation, water security problems are projected to intensify by 2030 in southern and eastern areas of Australia (IPCC, 2007: 9). The frequency of drought may increase by up to 20 per cent over most of Australia and climate change is expected to cause a higher incidence of water-borne diseases (DCC, Fact Sheet). Rainfall reductions in some areas of Australia are leading to lower water flows into rivers, wetlands and dams. In the Murray Darling, a 10 per cent change in rainfall has seen a 35 per cent reduction in stream flows (DCC, 2008a: 5). The Wet Tropics and Kakadu wetlands, alpine areas and deep coral reefs have been identified as particular areas at risk to the impacts of climate change (DCC, 2008b: 2–14). The Department of Climate Change and Energy Efficiency (DCCEE) has noted that by 2030, it is estimated that Australia will be exposed to a one degree Celsius increase in temperature, 20 per cent more months of drought, a 25 per cent increase in days of very high or extreme fire danger and increased storm surges and severe weather events.[6]

International responses to climate change

The Kyoto Protocol

The UNFCCC is a significant instrument in its own right as it contains the framework for an international regime to address climate change. It is also an important instrument because the Kyoto Protocol is a protocol to this convention. The Kyoto Protocol came into force on 16 February 2005 and sets emissions targets for 'developed' countries during the first commitment period (from 2008 to 2012) (*The Kyoto Protocol*, 1998). Parties who meet their targets are able to trade 'carbon credits' (reduction units) generated through any greater reduction of greenhouse gas emissions. Emissions trading is supplementary to domestic abatement action and is one of the 'flexible mechanisms' under the Protocol that parties can use to achieve their emission reduction targets.[7]

---

5   Recent research discussed in the Australian Government CPRS White Paper indicates sea levels may rise in some areas by up to 1.4 metres by 2100.

6   DCCEE, see: <http://www.climatechange.gov.au/climate-change/impacts.aspx> (accessed 12 August 2011).

7   Other flexible mechanisms include Joint Implementation (JI) projects and the Clean Development Mechanism (CDM). These flexible mechanisms enable developed country parties to the Protocol to implement projects in other developing country (CDM) or developed country (JI) parties. In order to obtain reduction units (or 'credits') from CDM or JI projects, methods of measuring emissions reduction must be verified

Under the Kyoto Protocol, greenhouse gas emissions reductions from forest 'sinks', which are essentially forests established on cleared land (reforestation), can be used to show compliance with binding emissions targets.[8] As discussed in more detail below, Indigenous peoples have argued that the inclusion of forest sinks and incentives for other large-scale 'clean' development in the Protocol may have detrimental effects on ecosystems and Indigenous livelihoods (*Declaration of Indigenous Peoples on Climate Change*, 2000; see also, *Declaration of the First International Indigenous Forum on Climate Change*, 2000).

Following the 2007 Conference of the Parties to the UNFCCC and members of the Kyoto Protocol in Bali, attention has focused on the ways in which benefits for avoided deforestation can be included in current and future mechanisms.[9] Avoided deforestation (including reduced emissions from deforestation and land degradation in developing countries (REDD)) may provide opportunities and a means of protecting forest-dependent communities from large-scale land clearing. However, there remain concerns with the design and implementation of avoided deforestation projects. In particular these concerns relate to how law and policy will be developed to enable appropriate tenure and natural resource (including carbon) rights and how benefits from avoided deforestation projects will be distributed to Indigenous peoples and forest dependent communities (IFIPCC, 2007; Graham-Harrison, 2007). While REDD projects are designed for implementation in developing countries, avoided deforestation projects have been accredited under voluntary carbon market standards in Australia.[10] Avoided deforestation projects are capable of attracting benefits (or credits) under the Australian Government's new Carbon Farming Initiative (CFI). As discussed below, tenure and access issues particular to Indigenous peoples are relevant to the participation in these regulatory proposals.

Indigenous land and water interests are threatened by not only the direct impacts of climate change but also the indirect impacts of national and international mitigation and adaptation responses. For example an agreement between the Ugandan Government and a multinational corporation to plant trees on 25,000 hectares of expropriated parkland led to the removal of residents from the area

---

and projects must demonstrate that the reduction in emissions is in addition to what would otherwise have occurred without the project (also called '*additionality*'). Various reduction units can be traded by parties to the Protocol and used to meet compliance targets. Australia, as a party to the Protocol, will be able to use the flexibility mechanisms to achieve its emissions targets. Seventh Conference of the Parties to the UNFCCC (COP 7) held at Marrakech 29 October – 19 November 2001, (hereafter Marrakech Accords).

8   Afforestation is the conversion of land that has not contained a forest for at least 50 years to forested land. Whereas reforestation by contrast is the conversion of land that was not forested on 31 December 1989 to forested land. In both cases forests must be human induced (planted by humans).

9   UNFCCC COP 13 Decision -/CP.13 (FCCC/CP/2007/6/Add.1. 14 March 2008): Reducing emissions from deforestation in developing countries: approaches to stimulate action. Agreed at COP 13/CMP 3, Bali, Indonesia, December 2007.

10   Avoided deforestation projects have been accredited under Australia's Greenhouse Friendly scheme. See: <http://www.climatechange.gov.au/greenhousefriendly/abatement/projects.html>

and restricted the income of locals from land use and grazing. Villagers were prevented from accessing the area to obtain food and traditional materials from the forest (Tauli-Corpuz and Lynge, 2008). Similarly, the identification by the IPCC that greenhouse gas emissions may be mitigated by the replacement of fossil fuels with biofuels has led to expansion of oil palm and other bioenergy crop plantations in Malaysia and Indonesia (Tauli-Corpuz and Lynge, 2008). Competing land uses not only create tensions and conflict with (and between) local communities, but also generate a struggle between food production and bioenergy cropping which in turn influences the supply of food markets. These activities can undermine the practical needs of Indigenous communities and infringe the fundamental human rights of peoples in these areas. Local communities are losing rights over land and facing increases in living expenses through, for example, an increase in the price of staple foods.

Indigenous peoples' participation in mitigation and adaptation strategies

As awareness about the potential impacts of climate change grew in the 1980s and 1990s, Indigenous peoples from around the world started to lobby for a role in the formulation or responses to predicted impacts of climate change. The 2000 Declaration of Indigenous Peoples on Climate Change expresses the integral nature of Indigenous peoples' rights in the management of natural carbon sinks on country in accordance with their culture, law, beliefs and use of these forests. The position articulated in the Declaration has been reiterated and built upon by Indigenous peoples at subsequent UNFCCC conferences and through the work of the United Nations Permanent Forum on Indigenous Issues.

Concerns articulated by Indigenous peoples in relation to international responses to climate change and the Kyoto Protocol include that:

- market incentives in relation to carbon sinks will lead to large-scale forest plantations and projects and a consequent loss of traditional country and abuse of ecosystems;

- discussions within the UNFCCC, along with practical implementation of the Kyoto Protocol, do not recognise the right of Indigenous peoples to adequate participation;

- measures to mitigate climate change are based on a worldview of territory that reduces forests, lands, seas and sacred sites to only their carbon absorbing capacity, and fail to take account of the fact that trees, vegetation and associated ecosystems are enmeshed with tangible and intangible cultural property rights;

- the importance of Indigenous peoples' traditional knowledge has not been adequately recognised in relation to climate change; and

- Indigenous communities have not been provided with sufficient information or resources to adequately respond to climate change. (*Declaration of Indigenous Peoples on Climate Change*, 2000: Articles 7–8; see also, Tauli-Corpuz and Lynge, 2008; *Declaration of the First International Indigenous Forum on Climate Change*, 2000)

Indigenous people have a 'special interest' in climate change and government responses to the impacts and effects of global warming. This interest is attributable not only to their unique physical and spiritual relationships with land, water and associated ecosystems (which gives rise to a particular vulnerability to climate change) but also to the specialised ecological and traditional knowledge they hold, which is relevant to finding 'best fit' solutions to climate change.

In 2008, the United Nations Permanent Forum on Indigenous Issues identified that a key barrier to the realisation of Indigenous peoples' adaptation capacities is the lack of recognition and promotion of their human rights. Many of these rights are reflected in the United Nations Declaration on the Rights of Indigenous Peoples (2007), which supports the participation of Indigenous peoples in climate change strategies and responses.[11] The United Nations Declaration on the Rights of Indigenous Peoples also recognises that respect for Indigenous knowledge, cultures and traditional practices contributes to sustainable and equitable development and proper management of the environment. The rights set out in the United Nations Declaration on the Rights of Indigenous Peoples are supported by other international instruments, including the Universal Declaration of Human Rights and other core human rights treaties (see, for example, *International Covenant on Civil and Political Rights,* 1966; *International Covenant on Economic, Social and Cultural Rights,* 1966). It is noted the Australian Government recently announced support for the United Nations Declaration on the Rights of Indigenous Peoples (Australian Labor Party, 3 April 2009). However, as identified by the United Nations Permanent Forum on Indigenous Issues statement welcoming Australia's endorsement of the Declaration, the

---

11 *UN Declaration on the Rights of Indigenous Peoples*, 2007. Further, the Declaration provides that Indigenous peoples:

- have the right not to be subjected to forced assimilation or destruction of their culture (Article 8(1));

- have the right to practice and revitalize their cultural traditions and customs. This includes the right to maintain, protect and develop the past, present and future manifestations of their cultures, such as archaeological and historical sites, artefacts, designs, ceremonies, technologies and visual and performing arts and literature (Article 11 (1));

- have the right to participate in decision-making in matters which would affect their rights (Article 18);

- have the right to maintain and strengthen their distinctive spiritual relationship with their traditionally owned or otherwise occupied and used lands, territories, waters and coastal seas and other resources (Article 25), *UN Declaration on the Rights of Indigenous Peoples.*

challenge ahead for Australia is how the contents of the Declaration will be implemented at a national level through the adoption of appropriate legislation and policies (Anaya *et al.*, 2009; Australian Human Rights Commission, 2010).

Several international and national legal instruments recognise the importance of Indigenous peoples' traditional knowledge in environmental management and biodiversity conservation (see, for example, *Universal Declaration on Human Rights*, 1948: Art 27; *International Covenant on Economic, Social and Cultural Rights*, 1966: Art 15; *International Labour Organisation Convention (No. 169) Concerning Indigenous and Tribal Peoples in Independent Countries*, 1992: Arts 15, 23; *Rio Declaration on Environment and Development*, 1992: Principle 22; see also, Dodson, 2007). The United Nations Convention on Biological Diversity (1992) promotes the importance of Indigenous peoples' traditional knowledge in the conservation and sustainable use of biodiversity.[12] Having ratified the United Nations Convention on Biological Diversity, Australia is under an obligation to (as far as possible and appropriate) respect, preserve and maintain the knowledge, innovations and practices of Indigenous peoples relevant to the conservation and sustainable use of biological diversity.[13] The United Nations Convention on Biological Diversity also promotes the wider application of traditional knowledge and practices, with the approval and involvement of the holders of such knowledge, and encourages the equitable sharing of benefits arising from the utilisation of such knowledge, innovations and practices. These concepts are important in the design of responses to climate change and Indigenous peoples' access to and participation in associated economic opportunities.

Other key principles of international environmental law further support Indigenous peoples' participation in strategies to address climate change (see, the Brundtland Report, 1987; see also, for discussion, Sands, 2003: 285–289).[14]

---

12   In particular Article 8(j): 'Each contracting Party shall, as far as possible and as appropriate: subject to national legislation, respect, preserve and maintain knowledge, innovations and practices of indigenous and local communities embodying traditional lifestyles relevant for the conservation and sustainable use of biological diversity and promote their wider application with the approval and involvement of the holders of such knowledge, innovations and practices and encourage the equitable sharing of the benefits arising from the utilization of such knowledge innovations and practices'.

13   Articles 8(j) and 10, *United Nations Convention on Biological Diversity*. The objects of the Commonwealth *Environment Protection and Biodiversity Conservation Act 1999* (Cth) (which embodies some of Australia's international obligations under the United Nations Convention on Biological Diversity) include the promotion of a partnership approach to environmental protection and biodiversity conservation through recognising and promoting Indigenous peoples' role in, and knowledge of, the conservation and ecologically sustainable use of biodiversity. See section 3 of the *Environment Protection and Biodiversity Conservation Act 1999* (Cth).

14   Principles such as: Ecologically sustainable development (ESD) or 'sustainable development (generally understood to be as articulated in the *Brundtland Report*: 'development that meets the needs of the present without compromising the ability of future generations to meet their own needs.'); Intergenerational equity (promoting conservation and sustainable use of biodiversity for the benefit of present and future generations); the precautionary principle (where there is a threat of significant reduction or loss of biodiversity, lack of full scientific certainty should not be used as a reason for postponing measures to avoid or minimise such a threat); and *the polluter pays principle* (the requirement that the costs of pollution should be borne by the person responsible for causing the pollution).

These principles are also relevant to the apportionment of responsibility in responding to the impacts of global warming. In particular, the principle of 'common but differentiated responsibility', derived from the concept of the common heritage of humankind and embodied in the UNFCCC, recognises historical differences in the contributions of different populations to global environmental problems, and the differences in their respective economic and technical capacities to address these problems (see, for discussion, CISDL, 2002). This principle encourages a shared response to climate change while protecting certain populations from a disproportionate burden in meeting mitigation and adaptation obligations (UNFCCC, 1992: Art 3; Sands, 2003: 286–289). The Clean Development Mechanism (CDM) is a mechanism through which countries can create tradable carbon credits by investing in projects in 'developing' countries. At the international level, this measure is an example of the principle's operation under the UNFCCC's Kyoto Protocol.

The principle of common but differentiated responsibility is usually interpreted at an international level, however it has application in a domestic setting, particularly in relation to the operation of the CFI and the design and implementation of the Australia Government's Clean Energy Future Plan (specifically, the proposed carbon pricing mechanism). Relatively speaking, it can be argued that non-Indigenous Australians have derived a greater economic benefit and played a greater role in creating environmental problems and should provide financial, technological, and other assistance to those who have contributed least to the creation of current environmental problems. Valuing the contributions of Indigenous peoples (past, present and future) is an essential component of equitable responses to climate change in Australia. For thousands of years Indigenous peoples have sustainably used and harvested country and their stewardship over biodiversity has sequestered significant volumes of carbon in soil, vegetation and trees. Employing a common but differentiated responsibilities approach to emissions trading and other responses to climate change in Australia promotes substantive equality between Indigenous and non-Indigenous peoples. Ignoring the inequalities between Indigenous and non-Indigenous peoples' capacities to adapt to climate change risks drifting into what Archbishop Desmond Tutu has described as a world of 'adaptation apartheid' (UNDP, 2007: 13).

The fundamental human right to self-determination, which includes the freedom and right to pursue economic, social and cultural development is embodied in many international legal instruments and is also relevant to responses to climate change (*International Covenant on Civil and Political Rights*, 1966; *International Covenant on Economic, Social and Cultural Rights*, 1996). Climate change policy design and mitigation strategies present an important opportunity to bring traditional knowledge and practices together with economic and social development. This opportunity will be diminished if governments 'mainstream' responses in a way that fails to accommodate the particular concerns and specialised interests of Indigenous peoples.

Historic policy documents (DCC, 2008b: 3–6) and Australia's status as a party to the UNFCCC, confirm Australia's commitment to the principle of common but differentiated responsibility and respective capabilities at an international level. However, despite the distinct socio-economic 'gap' between Indigenous and non-Indigenous Australians acknowledged by the Australian Government,[15] this principle has not filtered through to design of the voluntary carbon market and the compliance market at a domestic level. The basis on which Indigenous Australians are included in responses to climate change and the mechanisms by which disproportionate costs (Garnaut, 2008: 389; 2011)[16] and market access burdens will be minimised have not been articulated in the proposed scheme. Achieving 'real improvements and outcomes for Indigenous communities' under the Government's 'Close the Gap' policy agenda (FaHCSIA, 2009), including 'taking stock of the true extent of inequality' and 'forging corporate partnerships' necessarily involves ensuring legal foundations and policy incentives are present to grow opportunities and partnerships.

A domestic variation of the Kyoto Protocol CDM could be introduced as a design feature of the CFI or carbon pricing mechanism to implement the principle of 'common but differentiated responsibility' at a national level. Such a mechanism could draw from key aspects of the CDM; promote technology and capacity transfer and foster projects that deliver environmental, economic and social outcomes, as well as the maintenance or revitalisation of culture (Cardinoza, 2005: 197–210).

Measures to promote sustainable investment in Indigenous communities were absent in the previously proposed CPRS, (which barely addressed Indigenous involvement in land use, land use change and forestry activities). While the Australian Government has made an effort to include specific pathways for Indigenous peoples to participate in the CFI,[17] further legislative and systemic changes are needed to ensure access to meaningful economic development opportunities and provide incentives for investment.[18]

---

15   The Australian Government acknowledges that 'Closing the Gap' on the relative disadvantage facing Indigenous Australians is fundamentally important to building a fairer Australia. <http://www.fahcsia.gov. au/sa/indigenous/pubs/general/documents/closing_the_gap/closing_the_gap.pdf>

16   Costs of preparing projects for accreditation under compliance schemes or voluntary carbon offset standards and general costs of living – for example, as noted by Professor Garnaut in the Garnaut Climate Change Review Final Report in relation to the rising cost of fuel resulting from emissions trading: 'Remote Indigenous Communities in northern and central Australia are likely to be particularly affected, given their reliance on diesel fuel for power supply as well as transport.' (at p 389).

17   The Carbon Credits (Carbon Farming Initiative) Bill 2011 (Cth) proposes specific pathways for native title holders, which include: deeming provisions for a native title registered body corporate to be the project proponent in certain circumstances; confirmation that consents and other certificates from otherwise eligible interest holders (such as Crown land Ministers) are not required where a group holds exclusive possession native title; and expressly enables consents in relation to carbon offset projects to be provided in registered ILUAs.

18   The recently announced Clean Energy Future reforms include $22 million for an Indigenous Carbon Fund, which will assist Indigenous peoples involvement in the CFI. However, there remain a number of issues associated with State, Territory and Commonwealth legislative regimes to create, transfer and recognise carbon rights in country. These inconsistencies, as well as differing views of governments on the status of native title and other Indigenous tenure interests create uncertainty for Indigenous peoples involvement in the CFI and related markets.

Australia's response to climate change

A number of policy and legislative measures have been proposed in recent years to try and facilitate greenhouse gas emissions reduction and climate change mitigation and adaptation. The Rudd Labor Government attempted to introduce an emissions trading scheme in the form of the Carbon Pollution Reduction Scheme. More recently, the Gillard Labor Government has passed legislation to enable the CFI and proposed further reforms through the Clean Energy Future Plan. The Clean Energy Future Plan includes a fixed and floating phase emissions trading scheme known as the 'carbon pricing mechanism'.

In addition, the Australian Government has created and replaced a number of voluntary emissions reduction initiatives. Examples include the Greenhouse Friendly program and its replacement the National Carbon Offset Standard. The CFI creates a voluntary carbon market for emissions avoidance (reducing emissions to the atmosphere) and carbon sequestration (the storage of carbon in land, vegetation and trees) projects in the land use sector. CFI carbon credits will be capable of use to meet requirements under the new 'compliance' or compulsory emissions trading scheme (the carbon tax or 'carbon pricing mechanism').

The voluntary carbon market is relevant to companies and individuals that are not required by law to limit carbon emissions yet still wish to offset their greenhouse gas emissions. This behaviour may be driven by one or more of various imperatives including concern for environmental, marketing or branding considerations or to satisfy corporate social responsibility commitments. There are various standards and accreditation processes used to calculate emissions reduction for the voluntary market. A number of these standards and processes have been operating for many years, where as some, like the National Carbon Offset Standard or CFI are recent developments.

In order to have emissions reductions accredited under most voluntary markets, it is necessary to demonstrate that the project meets the following criteria:

- *Additionality* – abatement must be additional (beyond what would be undertaken as part of business-as-usual (common industry practice) or beyond what is required by law/regulation).
- *Permanence* – emissions reductions must be permanent. In the case of forest sinks, this requires that carbon remains stored and will not be released into the atmosphere in the future.
- *Measurability* – methodologies used to quantify the amount of carbon sequestered must be accepted and robust.
- *Transparency* – consumers and other interested stakeholders must be able to examine information on projects through the internet.

- *Independently verified* – eligibility of the project and the amount of carbon sequestered must be validated by an independent third party.

- *Registered* – units generated must be registered and tracked in a transparent and accessible registry. (DCC, 2008d: 15–16; *Carbon Credits (Carbon Farming Initiative) Act 2011* and associated Explanatory Memorandum)

These criteria are modelled on rules under the Kyoto Protocol for Clean Development Mechanism (CDM) projects. The CFI takes a slightly different approach to the assessment of 'additionality'. Under the CFI proponents must demonstrate a project passes the 'additionality test', which involves an assessment of legal additionality (whether the project or activities are required by law) and an assessment of whether the project is of a type included on a 'positive list'.[19]

There are cost implications in meeting these criteria, including consultant costs in designing and establishing projects and ongoing costs involved in monitoring and reporting on emissions reductions.

Tenure and 'additionality' requirements under these proposals complicate access to emerging markets for Indigenous peoples. Traditional lands are sometimes held in trust or not 'owned' by Indigenous peoples under a Torrens system of land tenure. Further, Indigenous peoples' lands often consist of conservation land; areas subject to reservations, declarations or covenants which require conservation and environmental management activities – such as joint management agreements and Indigenous Protected Areas. Demonstrating that revegetation, reforestation or land management activities in these areas satisfy additionality requirements (are in addition to what would otherwise occur) for the purposes of voluntary or compulsory markets is therefore difficult for Indigenous landowners and managers in these areas. To ensure access for these communities, definitions, 'lists' and requirements under new and proposed legislation need to accommodate this issue and the impacts of conditional land return on Indigenous land interests.

Further, the compulsory and voluntary drivers for sourcing carbon credits is likely to have a significant impact on the value of voluntary markets and on investment in voluntary offset projects. The priority for many liable entities under the proposed emissions tax or market will be investment in compliance under that scheme with secondary consideration of any additional and voluntary activities outside the scheme. For this reason, the final mechanisms which link

---

19 Section 41 of the CFI Act sets out the requirements that must be satisfied to pass the 'additionality test'. 'An offsets project passes the additionality test if the project is of a kind specified in the regulations (the positive list) *and* the project is not required to be carried out by or under a law of the Commonwealth, a State or a Territory' [emphasis added].

CFI projects to the carbon pricing mechanism will be relevant to Indigenous land holders and managers, as these links represent pathways to the highest value commercial markets.

Indigenous involvement and a greater opportunity for Australia

Issues relevant to the interaction between emerging carbon markets and Indigenous peoples in Australia include the following salient matters:

1. Indigenous peoples have unique cultural interests, economic development aspirations and legal rights and interests that must be respected, preserved and promoted where they intersect with carbon market opportunities;

2. Indigenous peoples possess many tangible and intangible assets that may be realised through meaningful and respectful partnerships and investment; and

3. As significant landholders, especially in northern Australia, the contribution of Indigenous peoples to mitigation efforts need to be recognised as a major component of the national mitigation response.

While there have been attempts to engage with and consult Indigenous peoples in relation to policy announcements and regulatory proposals more recently, meaningful engagement with, and analysis of, these issues is yet to be seen by the Australian Government. In contrast the New Zealand Government specifically examined the potential impacts (positive and negative) of an emissions trading scheme on the interests of Māori (Insley and Meade, 2008). This study, undertaken by the New Zealand Government with 37 Degrees South Limited and Cognitus Advisory Services Limited, was preceded by general consultation with Māori in relation to climate change, its impacts and the Kyoto Protocol (DPMC, 2001). The study was designed to inform the consultation processes of the New Zealand Government with Māori and the finalisation of the Government's climate change policy (Insley and Meade, 2008). The final report on the relative impacts of a trading scheme indicates that Māori face increased burdens and restricted opportunities under an emissions trading scheme, unless concessions and exemptions are made (Insley and Meade, 2008). While the magnitude of impacts depend on particular elements of the New Zealand scheme, Māori are more likely to be affected by increased household electricity and fuel prices, more likely to be exposed to increased costs and burdens in the fishing sector (including adverse employment impacts) and parts of the forestry sector. Also, the ability of Māori to mitigate any disproportionately negative impacts or take advantage of Kyoto forestry activities will be constrained relative to non-Māori due to transaction costs and trading scheme related penalties and land use restrictions (Insley and Meade, 2008).

This type of evaluation is critical to the principle of fairness in the Australian Government's assessment of design options for the 'compliance' market (or carbon tax). Meaningful and ongoing engagement with Indigenous communities in climate change responses should involve an examination of the foreseeable impacts (positive and negative) of emissions trading on Indigenous peoples' land use, development opportunities and living expenses. Such an approach is also supported by recommendations of the United Nations Permanent Forum on Indigenous Issues (Tauli-Corpuz and Lynge, 2008).[20]

Climate change, carbon rights and the interests of Indigenous peoples in land and waters

Climate change related laws, regulations and markets have the potential to further decrease or limit Indigenous peoples' rights and interests in country and its resources through the extinguishment or suspension of native title or by restricting rights in relation to the access and use of land and resources. The progressive 'unbundling' of conventional property interests through legislation (for example separating rights and interests in water and carbon from land ownership) creates a regime for the piecemeal appropriation of traditional land and resources.

Most Australian States and Territories have legislated to provide a basis for the legal recognition of carbon rights in trees and natural resource products. However the nature of these carbon rights varies across jurisdictions. There is inconsistency in relation to the land on which carbon rights may be created (private or public or Crown land), whether they create an interest in land (or constitute a new and separate property interest) (DEH, 2005: Ch 2; see also, Hepburn, 2008), and whether harvesting rights are separate from sequestration rights (Peel, 2007: 90). In addition, many State regimes protect a carbon interest holder's rights by registering an instrument on the relevant land title. Registration of interests on titles presents an obstacle for many Indigenous people in Australia as their interests in land often prohibit or restrict the creation of third party interests or require the consent of relevant government ministers to do so, or the tenure interests held by Indigenous communities cannot be registered on the Torrens system. The nature and effect of carbon rights creates a complex interaction with other legal interests, including native title.[21]

---

20 For example, a recommendation that the business community and its regulators should incorporate Indigenous peoples' rights into their plans for economic development, as stakeholders, land rights holders and on a human rights basis.

21 See generally for example: *Conveyancing Act 1919* (NSW) as amended by the *Carbon Rights Legislation Amendment Act 1998* (NSW); *Forestry Act 1959* (Qld) as amended by the *Forestry and Land Title Amendment Act 2001* (Qld); *Forest Property Act 2000* (SA); *Climate Change Act 2010* (Vic); *Carbon Rights Act 2003* (WA); *Forestry Rights Registration Act 1990* (Tas) as amended by the *Forestry Rights Registration Amendment Act 2002* (Tas).

The use of market mechanisms to address contemporary environmental issues and the requirements for participation in these markets (for example, demonstrating 'additionality', as discussed above) highlight the problems with historic and current conditional land return practices of Australian governments (for example, handback-leaseback arrangements, dedicated conservation areas and conditional tenure interests). This practice impairs the ability of Indigenous peoples to participate in economic development opportunities, a suggestion that is not new to consideration of Indigenous peoples rights in Australia. In 1974, Justice Woodward noted that the provision of adequate and meaningful rights to use and develop land was one way in which to achieve economic development in Indigenous communities (Woodward, 1974). It is essential to Indigenous participation in emerging environmental market opportunities that the statutory and policy infrastructure supporting these markets respect the integrity and comprehensiveness of Indigenous rights and interests in land and water and the rights of Indigenous peoples to derive contemporary benefits from their rights and responsibilities.

Native title

When Britain asserted sovereignty over Australia it also acquired 'radical title' to land in Australia (*Mabo v Queensland [No 2]*, 1992). However, as held by the High Court of Australia, Indigenous law with respect to land and waters survived the Crown's acquisition of sovereignty and 'radical title' (*Mabo v Queensland [No 2]*, 1992). Radical title alone is 'merely' a logical assumption needed to give support to the Crown's power to grant an interest in land to others or appropriate to itself ownership of areas within its territory.[22] Radical title does not of itself confer full and beneficial ownership of land and waters to the Crown; it is qualified or 'burdened' by native title, which is the name given to the rights and interests arising under Indigenous law and custom that are recognised by the common law (*Mabo v Queensland [No 2]*, 1992: 89, 94, per Brennan CJ; *Wik Peoples v Queensland*, 1996: 128, per Toohey J; see also, Bartlett, 2004: 205). As observed by Brennan CJ in *Mabo v Queensland [No 2]*, 1992:

> the common law of this country would perpetuate injustice if it were to continue to embrace the enlarged notion of terra nullius and to persist in characterizing the indigenous inhabitants of the Australian colonies as people too low in the scale of social organization to be acknowledged as possessing rights and interests in land. Moreover, to reject the theory that the Crown acquired absolute beneficial ownership of land is to bring the law into conformity with Australian history. The dispossession of

---

22   *Mabo v State of Queensland [No 2]* (1992) 175 CLR 1, per Brennan CJ at [52] and [53].

the indigenous inhabitants of Australia was not worked by a transfer of beneficial ownership when sovereignty was acquired by the Crown, but by the recurrent exercise of a paramount power to exclude the indigenous inhabitants from their traditional lands as colonial settlement expanded and land was granted to the colonists. Dispossession is attributable not to a failure of native title to survive the acquisition of sovereignty, but to its subsequent extinction by a paramount power. (*Mabo v Queensland [No 2]*, (1992) 175 CLR 1, per Brennan CJ at [63])

As such, areas of unalienated Crown land (which could be described as areas of unallocated Crown radical title) remain burdened by native title rights and interests. However, while the rights and privileges conferred by native title were unaffected by the Crown's acquisition of radical title, the acquisition of sovereignty exposed native title to extinguishment by a valid exercise of sovereign power inconsistent with the continued right to enjoy native title.[23]

The content of native title is determined by the laws and customs governing the relationship of an Indigenous group to the country to which it is connected. However, native title is subject to several limitations, some of which are outlined here. First, with a couple of specific exceptions, native title rights and interests have not been found to include a commercial right to trade.[24] Second, the acquisition of sovereignty exposed native title to 'extinguishment' by the valid exercise of sovereign power (in a manner inconsistent with the continued right to enjoy native title) (*Mabo v Queensland*, 1992; see also, *Native Title Act 1993* (Cth) (NTA)). As a result, subject to protections in the NTA, the Crown can validly grant interests in land that are wholly or partially inconsistent with (and override) native title. Third, the laws and customs giving rise to the native title must be traditional, that is, there must be a clear nexus between the contemporary and pre-sovereign societal system (*Mabo v Queensland*, 1992; *Members of the Yorta Yorta Aboriginal Community v Victoria*, 1998). The effects of these limitations on native title holders' participation in measures addressing climate change is addressed below.

Whether the scope of native title includes the right to carbon (or carbon rights) is the subject of debate between governments and Indigenous communities. This debate is likely to intensify as the value of carbon increases. The following discussion explores the nature of native title and scope of potential commercial or trade interests protected by (or associated with) its recognition.

---

23 See: *Mabo v State of Queensland [No. 2]* (1992) 175 CLR 1, per Brennan CJ at [51], [62], [63], [66], [67], and [83].

24 It is noted that consent determinations in relation to land in the Torres Strait have recognised commercial native title rights and interests. Further the recent decision in *Akiba v Queensland [2010]* FCA 643 recognised the native title right to trade (this decision was subsequently appealed and at the time of writing the decision remains reserved).

Native title right to trade

Native title is recognised as bundle of rights and interests in relation to land (*Western Australia v Ward*, 2002). The 'inherently fragile' (*Fejo v Northern Territory*, 1998: 105, per Kirby J) nature of native title is likely to restrict the participation of native title holders in carbon markets, particularly where participation requires the creation (and registration) of third party interests in trees or other natural resources.

While exchange and sharing of resources has been demonstrated in the context of native title, there has been a reluctance to recognise a native title right to trade (with the exception of recognition of native title rights and interests in the Torres Strait which may be validly exercised for commercial purposes) (see, for example, *Saibai People v Queensland*, 1999; *Kaurareg People v Queensland*, 2001; *Mabuiag People v Queensland, 2000; Masig People v Queensland*, 2000). However, a right to trade may be recognised as part of exclusive rights to use and enjoy land and waters (*Commonwealth v Yarmirr*, 1999: 250, per Beaumont J and Von Doussa J).

The recent decision in *Akiba on behalf of the Torres Strait Islanders of the Regional Seas Claim Group v Queensland (No 2)* ([2010] FCA 643) provides an example of an actively litigated outcome where the right to trade, or commercial use of native title rights and interests has been recognised. It is noted that, at the time of writing, an appeal against the decision at first instance has been heard and judgement is reserved. In the initial Federal Court decision, Justice Finn found that it was

> by no means apparent ... at least in relation to the sea – and particularly in waters with the abundant resources Torres Strait has ... absent a legislative regime to the contrary, why marine resources may not be exploited by those who care to do so for trading and commercial purposes, though they lack entirely any exclusive right to possession of the area or do not purport to assert any such right. (*Akiba* [528]–[529])

His Honour accepted that the evidence established that the Islanders sold marine resources for money and that 'the fundamental resource-related right of use was the right to take. Use of what was taken was unconstrained, save by considerations of respect, conservation and the avoidance of waste' (*Akiba* [528]–[529]).

In *Northern Territory v Alyawarr, Kaytetye, Warumungu, Wakaya Native Title Claim Group* (2005: 157), the Court found sufficient evidence to support the right to share or exchange subsistence and other traditional resources obtained on or from the land and waters. However, the Court did not find sufficient evidence to trade in the resources of the area. In this case the Northern Territory argued that the evidence of trade presented by the applicant made no reference to commercial or profit motives or any level of business operation.

As contended by Langton and others (2006a), such a viewpoint neglects the distinct nature of Indigenous peoples' transactions and economic relations and ignores the inherent agency of resources as commodities with multiple meanings and value. It may be argued that a contemporary expression or adaptation of a right of exchange (where evidenced) includes the exchange of resources for money. In the same way that Indigenous communities adapt to contextualise and normalise interactions with new discoveries and foreigners, adaptation of law and custom could clearly accommodate interaction with cash economies. The general absence of contemporary recognition of traditional economies and commercial rights also limits the use of native title as a means of underwriting economic enterprise.[25]

Even where native title right to trade is recognised the susceptibility of native title to regulation or extinguishment may undermine its use as the sole means of accessing emerging environmental markets.[26]

One new and significant market, the CFI, deals with these issues in part by deeming 'exclusive possession' native title interest holders to also hold the relevant carbon rights needed to generate tradeable carbon credit units. Also, the CFI confirms that responsibility for a carbon sequestration project can be transferred (or consented to) by a native title holding group through a registered Indigenous land use agreement. Measures like these are an efficient and effective way of addressing certain obstacles to Indigenous participation in carbon markets. The alternative, forcing parties to go through extensive and costly 'proof' exercises to confirm the requisite carbon right, only delays potential projects and acts as a disincentive to potential project partners.

25 It is noted that consent determinations in the Torres Strait expressly include economic purposes in recognition of native title rights to conserve, use and enjoy the natural resources of the determination area for social, cultural, economic, religious, spiritual, customary and traditional purposes (see, for example: *Saibai People v Queensland* [1999] FCA 158, *Kaurareg People v Queensland* [2001] FCA 657 (23 May 2001), *Mabuiag People v Queensland* [2000] FCA 1065 (6 July 2000) and *Masig People v Queensland* [2000] FCA 1067 (7 July 2000)).

26 As considered by Justices Beaumont and Von Doussa in *Commonwealth v Yarmirr* [1999] FCA 1668: 'any final consideration of a claim to a right to fish, hunt and gather within these waters for the purposes of trade would need to take into account the impact of the relevant respective fishing legislative regimes of South Australia, the Northern Territory and the Commonwealth, the various forms of applicable fisheries legislation and administrative action there-under, which clearly had at least the potential to affect a claim by any person to fish or hunt in these waters, were summarised by the judge (at 594–599) ... it will suffice for us to say that, by this means, any right of the public to fish for commercial purposes, and any such traditional right, were at least regulated and possibly wholly or partly extinguished by statute or executive act, or both' (at 255).

Native title and incidental commercial benefits

In addition to the possible recognition of a native title right to trade, there is scope for legal recognition and preservation of contemporary economic interests deriving from native title. The issue of whether the adaptation of a traditional practice, attracting an economic benefit, means that the practice is no longer considered to be 'traditional' was considered in *Neowarra v State of Western Australia* (2003). In addressing the question whether the use of canvas and the sale of artworks to tourists is consistent with tradition, Justice Sundberg viewed the sale of artworks as an 'incidental spin off'. His Honour accepted the rationale for developing painting on canvas (to educate children about their heritage) and considered the practice to be 'traditional' in the sense of section 223(1) of the NTA.[27] Importantly, Justice Sundberg further accepted that the practice does not lose its 'traditional' character because it has an incidental economic advantage (*Neowarra v State of Western Australia*, 2003: 341).[28]

Extending Justice Sundberg's reasoning to other traditional activities, it may be argued that native title can support the economic use of traditional rights, for example, the maintaining of and caring for country in a manner which provides an incidental economic advantage. As such, native title may provide an opportunity for participation in carbon markets through carbon offset and abatement projects and managing country.

Such an extrapolation is relevant for example in the context of 'patch' burning of the Martu people in the north-western section of the Western Desert. The Martu People are native title holders in this area of Western Australia (*Martu People v State of Western Australia*, 2002). As discussed by the former Desert Knowledge CRC and others (Campbell *et al.*, 2007: 9; also cited by, Bird, *et al.*, 2005: 443–463), Martu women undertake burning activity, which reveals the tracks and dens of small burrowing animals and improves hunting efficiency (Campbell *et al.*, 2007). The burnt areas resulting from the women's use of fire have a collateral benefit of mitigating wild fires and sustaining biodiversity (Campbell *et al.*, 2007). The minimisation of wild fires preserves vegetation and increases the capacity of the ecosystem to maintain carbon sequestration

---

27 Section 223 of the NTA sets out the meaning of native title or native title rights and interests. Native title claimants must satisfy this definition in order to have native title recognised.

28 At [341]: 'I turn to the suggestion that the painting of artworks may not be traditional because they are sold to tourists. At sovereignty the claimants' ancestors painted on rock surfaces and renovated the paintings either annually or as required. While some renovation is still carried out, the remoteness of many Aboriginal people from their Wanjina sites prompted Donny Woolagoodja and his Wanjina Corporation to keep up the painting tradition by encouraging people to paint on canvas so as to educate the children about their heritage. The sale to tourists and others of the works is an incidental spin off. Once it is accepted, as I do, that the rationale for the development of painting on canvas at Mowanjum is that given by Donny Woolagoodja and other artists such as Mabel King, the practice is "traditional" in the sense of that word in s 223(1), and does not lose that character because it has an incidental economic advantage.'

(Campbell *et al.*, 2007). As such, it could be argued that the commercialisation of these activities is a beneficial incidental 'spin off' to native title rights and interests.

Adaptation of traditional laws and customs is relevant in considering protections under the NTA afforded to traditional practices that, incidental or otherwise, mitigate the effects and impacts of climate change.

While native title claimants must demonstrate a clear nexus between the contemporary and pre-sovereign societal system, adaptation and change of laws and customs over time is not fatal to native title (*Members of the Yorta Yorta Aboriginal Community v Victoria*, 2002). In assessing the significance of change to, or adaptation of, traditional law and custom, it is necessary to determine whether the change or adaptation is of such a kind that it can no longer be said that the rights and interests asserted are possessed under the traditional laws acknowledged and traditional customs observed by the relevant peoples (*Members of the Yorta Yorta Aboriginal Community v Victoria*, 2002: 83 per Justices Gleeson, Gummow and Hayne).

Where native title is demonstrated, section 211 of the NTA may operate to protect native title holders from certain attempts to regulate the exercising their native title rights and interests.[29] Despite its positive application, section 211 does not preserve a 'right to trade', as subsection 211(2)(a) stipulates that the preservation of native title rights applies only for the purposes of satisfying personal, domestic or non-commercial communal needs. However, while section 211(2)(a) may limit the preservation of a right to trade or activities carried out for the purposes of commercial benefit, on the basis of the reasoning in *Neowarra* it may be possible for native title holders to enjoy protection under section 211 for practices that are carried out primarily for personal, domestic or non-commercial communal needs yet which attract an indirect and incidental commercial benefit.[30]

---

29   Section 211 of the NTA provides for the preservation of certain native title rights and interests where such a right or interest may be otherwise restricted by Australian statute. Effectively, the section provides that where a law prohibits or restricts persons from certain activities other than in accordance with a licence, permit or other instrument, it does not prohibit or restrict the pursuit of that activity in certain circumstances where native title exists and native title rights are exercised for personal, domestic or non-commercial needs. See *Yanner v Eaton (1999)* 201 CLR 351 in which the High Court found that section 211(2) of the NTA prevented Queensland legislative provisions from prohibiting a native title holder from exercising his native title right to hunt for crocodiles.

30   For example traditional burning practices for hunting personal, domestic or non-commercial communal purposes that carry an indirect and incidental biodiversity conservation and fire mitigation benefit *Yanner v Eaton* (1999) 201 CLR 351.

Native title future act regime considerations

The native title future act regime also presents a way in which Indigenous groups may use existing agreement-making mechanisms to participate in carbon offset projects. Participation may be facilitated through negotiated agreements for use and development projects on country (for example through an Indigenous Land Use Agreement). Such negotiations may provide a basis for arrangements under which companies seek to offset carbon emissions for their business or project by supporting Indigenous peoples' rights to care for country. Investment and partnering in such local enterprise may foster knowledge and skill exchanges and help proponents achieve corporate social responsibility and regulatory compliance objectives.[31] However, as noted above, this investment will depend on projects being able to meet relevant eligibility requirements under proposed schemes, which, will be difficult if they are implemented as drafted.

Agreements involving large-scale infrastructure, energy or mining projects are perhaps the most fertile area for negotiating carbon benefits, especially for projects resulting in, or facilitating, significant increases in greenhouse gas emissions. In particular, mining projects provide a specific opportunity to negotiate participation in carbon offset projects due to the statutory requirement to negotiate under the NTA. Looking for ways to expand agreements in this way is consistent with current policy and reform objectives of the Australian Government to optimise benefits from native title agreements.[32]

As set out above, unalienated Crown is burdened by native title (*Mabo v Queensland*, 1992). As a result, trees and vegetation, as fixtures of land, are also 'burdened' by native title rights and interests where these rights and interests have not been extinguished by the valid grant of an inconsistent interest. The trend toward legislation enabling governments to create rights and trade carbon stored in trees and soil on Crown land assumes that the Crown has an unencumbered absolute beneficial interest in the underlying Crown land. However, as a burden on the Crown's radical title and where it has not been extinguished, native title (and the operation of the NTA and *Racial Discrimination Act 1975* (Cth)) limits the manner in which governments can deal with interests in Crown land.

---

31 Such as conditions designed to achieve social or environmental outcomes and imposed on proponents as part of planning and environmental approvals processes.
32 At the time of writing, the Australian Government is developing an Indigenous Economic Development Strategy which involves examining ways to improve economic development outcomes for Indigenous people. This initiative also forms part of the Australian Government's broader policy agenda to 'close the gap' between Indigenous and non-Indigenous Australians in key areas of Indigenous disadvantage. The Government released a discussion paper in December 2008, available at: <http://www.fahcsia.gov.au/sa/indigenous/progserv/land/Documents/native_title_discussion_paper/default.htm>

In general, carbon markets (or standards used to accredit offsets generated in certain ways) require a carbon sequestration project proponent to demonstrate that they hold the requisite carbon right (right to benefit from carbon stored in the land). As mentioned above, in order to facilitate the creation and transfer (and trade) of 'carbon' Australian States have enacted legislative frameworks to confirm recognition of carbon rights.

There is unlikely to be an issue where a State or Territory government grants carbon rights to another party where native title has been extinguished (and therefore the grant cannot affect native title rights and interests). However, in circumstances where native title rights and interests have not been extinguished, the grant or creation of carbon rights (and/or the access arrangements needed for a project to exploit these carbon rights) is likely to attract procedural and substantive rights under the NTA future act regime. Where such a future act is valid under the NTA, compensation may be payable for any impairment of the use and enjoyment of native title rights and interests. This affords some protection to native title holders and registered claimants, which is particularly important given the significant consequences statutory schemes like the CFI allow: the ability for a regulator to impose ongoing land management obligations and restrictions on areas used for projects where the proponent (who may not be the landowner/holder) fails to comply with their obligations.[33]

As such, the potential use of land for carbon offsets is also relevant to negotiations permitting a decrease in traditional owner access or use of country. Lost opportunity or income as a result of an alternative land use may be relevant to compensation arrangements in these circumstances. While quantification of native title is a difficult and relatively unprecedented area, over time estimates of lost income from potential carbon abatement activities may be made by reference to market value of projects and carbon yields.

Land rights legislation, specific freehold grants and Indigenous Land Corporation acquisitions

Many States and Territories have enacted land rights legislation that provides for grants of communal freehold land to Indigenous groups. Further, land grants have been made to Indigenous peoples through native title settlements or specific legislation.[34] Land grants typically involve inalienable freehold land

---

33   See for example the operation of the CFI 'carbon maintenance obligation' (Carbon Credits (Carbon Farming Initiative) Bill 2011, Part 8).

34   See the settlement agreement package for the Wotjobaluk, Jaadwa, Jadawadjali, Wergaia and Jupagalk Peoples Application for determination of native title in Victoria in which freehold title to certain parcels of land was transferred back to the traditional owners; *Wotjobaluk, Jaadwa, Jadawadjali, Wergaia and Jupagalk*

which is held on trust for the benefit of a group. Alternatively, legislation enables the reservation of land for the use and benefit of Indigenous peoples. The Indigenous Land Corporation (ILC) also provides an avenue for Indigenous groups to acquire land in Australia.[35]

More certain land tenure generally provides greater scope to use land for economic development. In contrast to native title holders, Indigenous groups holding freehold land enjoy greater security of tenure which may be used as a platform for direct participation in environmental markets.[36] Full realisation of these interests hinges on the extent to which statutory forms of Indigenous land are recognised under subsequent regulatory and development legislation. This issue is live in relation to tenure requirements for carbon sequestration projects under existing and proposed legislation.

Economic development opportunities

Indigenous peoples in Australia have long performed activities which generate commodity and non-commodity services (for all Australians) from the natural environment (Campbell *et al.*, 2007). Many environmental services performed by Indigenous peoples are not 'new' to Federal, State and Territory governments. Government departments and agencies have been involved in joint and cooperative management arrangements with Indigenous peoples for some time. However, the current threats of climate change and associated 'low carbon' context significantly reinforces the need to more appropriately value these services and provide adequate financial and regulatory infrastructure to enable access to, and growth of, new opportunities.

The West Arnhem Land Fire Abatement Project (WALFA Project) case study is an example of linking traditional practices and knowledge with economic benefits from climate related opportunities. It illustrates the tangible and intangible assets of Indigenous communities that may be realised through

---

native title determinations: what they mean for the Wimmera region, 2005. See also the Agreements, Treaties and Negotiated Settlements database, Indigenous Studies Program, University of Melbourne, available at: <http://www.atns.net.au/agreement.asp?EntityID=3126> For examples of specific legislative land grants see: *Aboriginal Land (Manatunga Land) Act 1992* (Vic); *Aboriginal Lands Act 1991* (Vic).

35 Further information about the Indigenous Land Corporation is available at: <http://www.ilc.gov.au/site/page.cfm>

36 In the same way as farmers and other private land owners are deciding to take advantage of incentives and payments for changes in land use and management, Indigenous land-owners may be able to increase carbon uptake through revegetation, cultivation of soil and other land management practices. The federal government's 'Working on Country' initiative is an example of such a program designed for Indigenous land holders (although it is noted that long term leasing arrangements in the Northern Territory may add risk to tenure security – and therefore investor confidence – for these purposes). Also, the recent *Caring for Our Country* initiative of the federal government proposes funding to assist Indigenous peoples enter the carbon market: <http://www.nrm.gov.au/funding/future.html>

meaningful and respectful partnerships and investment (Tropical Savannas Cooperative Research Centre, see: <http://www.savanna.org.au/savanna_web/ information/arnhem_fire_project.html> accessed 31 October 2008; Aboriginal and Torres Strait Islander Social Justice Commissioner, 2007: Ch 12; Garnaut, 2008: 557). The WALFA Project is a carbon offset project in western Arnhem Land in the Northern Territory, formed to implement strategic fire management for the purposes of offsetting greenhouse gas emissions from a Liquefied Natural Gas plant in Darwin Harbour. The WALFA Project reduces greenhouse gas emissions by adapting traditional fire management practices in areas that are prone to unchecked wildfires. The project has direct and collateral ecological benefits, by reducing net greenhouse gas emissions from wildfire and by conserving environmental and cultural values in the adjacent World Heritage-listed Kakadu National Park (Aboriginal and Torres Strait Islander Social Justice Commissioner, 2007: Ch 12). Over the first four years of the WALFA Project, fire management has abated equivalent to about 488,000 tonnes of CO_2, which exceeds the 100,000 tonnes per year contemplated by the Agreement (see: <http://savanna.cdu.edu.au/information/arnhem_fire_project.html>).

Case study research undertaken by the CSIRO confirms potential commercial opportunities in relation to carbon sequestration on Indigenous lands (Heckbert et al., 2008). Using a price estimate of $20/tonne CO_2 equivalent sequestered, a case study involving fire management on an ILC property in the Northern Territory generated an estimated annual income of $208,000 (Heckbert et al., 2008). Similarly, a forestry case study on a station in Queensland generated a possible yearly revenue at over $200,000 (again assuming a carbon price of $20/tonne CO_2 equivalent). While these projects present exciting opportunities for those who chose to participate, care must be taken in the design and implementation phases, particularly for projects drawing on Indigenous understandings about country, to ensure adequate and appropriate protection of cultural protocols, confidentiality and traditional knowledge.

In addition to the potential for Indigenous peoples to participate in carbon related markets through involvement in land use and development projects, scope also exists for other opportunities through collaborative projects relating to climate change and environmental management, which support and/or foster shared understandings about country (for example existing caring for country programs and the Indigenous weather knowledge project) (DEH, 2004; Climate Change Research Centre, 'Sharing Knowledge', see: <http://www. sharingknowledge.net.au/>; Bureau of Meteorology, 'Indigenous Weather Knowledge' program, <http://www.bom.gov.au/>).[37]

---

37   The need to adequately protect traditional knowledge and cultural property in relation to these projects is noted.

The Victorian Government's Land and Biodiversity at a time of Climate Change 'Green Paper – White Paper' consultation process prioritises the knowledge, skills and perspectives of Indigenous communities and suggests they should inform land and biodiversity management decisions (DSE, 2008, 2009). The Victorian Government is also examining access and benefit sharing arrangements in relation to the use of Indigenous traditional knowledge and ways of enabling greater involvement of Indigenous peoples in land management (DSE, 2008).[38] These initiatives are a positive step toward realising meaningful development opportunities for Indigenous peoples.

The future of agreement making and meaningful participation

Despite the opportunities for Indigenous peoples to increase their involvement in environmental markets, there are significant access issues in relation to emerging opportunities. These issues centre around financial and human resources, land tenure, water rights and intellectual property (protection of traditional knowledge). In addition, there are commercial risks and potential liabilities that come with many of these opportunities.

Joint and cooperative management arrangements

It is important that legal and policy infrastructure supports Indigenous peoples' access to and participation in environmental markets by recognising the nature of existing Indigenous tenure and land management arrangements and allowing for these interests to grow. Established joint-management agreements in a number of States and Territories provide a basis on which to create underlying contracts and agreements for environmental services and other offset projects. However as mentioned above, these methods of conditional land return will need to be examined by Indigenous parties on a case-by-case basis to assess whether they provide the requisite rights and interests to participate in economic development opportunities such as environmental (carbon) markets.

As mentioned, carbon markets highlight significant issues associated with the historic and current practices of governments for settlement of native title claims and other transactions with Indigenous peoples. In many cases the 'additionality' hurdle means that potential project areas will need to be free

---

38   It is noted that requirements for access and benefit sharing agreements exist under the Environment Protection and Biodiversity Conservation Regulations 2000 (for biological resources taken from Commonwealth areas). A benefit-sharing agreement under Part 8A of the Regulations must provide for reasonable benefit-sharing arrangements, including protection for, recognition of and valuing of any Indigenous people's knowledge to be used.

of specific funding conditions, conservation covenants and various lease back requirements that may confuse ultimate management rights and responsibilities of participants.

Depending on the underlying tenure arrangements, joint management arrangement (involving the transfer of freehold title to an Indigenous group) may enable Indigenous peoples, as landowners and joint-land managers to develop appropriate management plans and enjoy benefits from carbon sequestration through land use management and forestry activities. The position is less clear for co-operative land management arrangements (which generally do not involve a transfer of title). The ownership of underlying land has relevance in the creation of registerable carbon and forestry rights under State and Territory legislation. Issues of land tenure are critical to the design of carbon projects under the existing and proposed schemes as the involvement of Indigenous peoples can be significantly restricted depending on the nature of their tenure interests across jurisdictions.

Meaningful rights and access issues

Indigenous peoples' access to emerging carbon market opportunities will be significantly and disproportionately affected by less obvious matters such as; access to finance and investment, secure water flows (water rights) and adapting to changing landscapes and country due to the impacts of climate change.

A common issue identified in relation to both the WALFA Project and Martu examples is the vulnerability of these projects to changes in policy and support structures. In particular the centralisation of essential infrastructure away from small remote outstations in favour of larger settlements emphasises the need for meaningful rights, particularly land rights and water rights, to facilitate engagement in development opportunities (see, Campbell, *et al.*, 2007; Gerritsen, 2007). This engagement presents an opportunity for Indigenous peoples to support their independent activities and practices, which continue despite changing policy and frameworks. Sustaining these practices and activities is essential to sustaining culture and identity, and providing pathways for support that are independent of changing government policy can assist communities to determine their own futures on their own terms and aspirations.

Maximising Indigenous development opportunities will require governments to revisit the fundamental nature of 'primary' Indigenous tenure rights and the recognition of existing tenure arrangements through 'secondary' schemes. Where ownership of, or interests in, land and waters are granted to Indigenous peoples, these interests must be recognised as a basis on which to participate new opportunities emerging through subsequent policy and legislation.

Access to emerging opportunities may be limited by inadequate financial assistance or related incentives for investment in Indigenous owned, managed or partnered projects. Carbon projects take time, and require financial and human resources (Gerritsen, 2007).[39] All government departments involved in environmental and water management, climate change and Indigenous affairs must collaborate to find ways to promote awareness and establish the foundations necessary to access emerging opportunities. In addition, consultation processes in relation to new and important government initiatives must be culturally appropriate. Detailed and lengthy policy documents, with short consultation timeframes, detailing important opportunities and risks is not an adequate way in which to engage and consult with the majority of Indigenous stakeholders.

The CPRS White Paper noted that water security is a major challenge in southern parts of Australia and that the costs of meeting this challenge will be significant (DCC, 2008b: 2–11). Further, it is noted in this report that stream flows in the Murray Darling Basin could fall by nearly 50 per cent by the end of the century. If such a prediction eventuates, it would obviously lead to significant limitations and impacts on cropping, aquatic ecosystems, terrestrial ecosystems and local and farming communities (Garnaut, 2008, 2011).

Inherent in environmental and ecosystem changes are changes and adaptation of Indigenous peoples' practices, landscapes and existence. According to the Murray Lower Darling Rivers Indigenous Nations, there are approximately 75,000 Indigenous people in the Murray Darling Basin and most of these people are traditional owners belonging to about 40 Indigenous Nations (See MLDRIN website, <http://www.mldrin.org.au/about/> accessed 20 August 2011). The ecological, agricultural and economic significance of the Murray Darling Basin is well known and documented. The impacts of altered stream flows, salinity, fish stocks and associated land and water ecosystems affects traditional owner use and enjoyment of cultural rights and interests in this region. Access to adequate water supplies to support traditional and cultural activities will also affect the degree to which Indigenous peoples can establish appropriate and viable economic and cultural projects.

Alternatives to agreement making and opportunity

Indigenous peoples represent a particularly vulnerable population in relation to climate change and related damage. For some time concern has been expressed about the serious health and lifestyle impacts of climate change on Indigenous

---

39 For example the WALFA Project took a number of years to eventuate. The exercise was initiated in 2005 and designed over a two year period (2006/07), following five years of preliminary data gathering and fieldwork in 2000–04.

peoples (*The Albuquerque Declaration*, 1998).[40] Changes in temperature and the environment are forcing Indigenous peoples to adjust strategies of hunting, fishing and travel, causing interference with residence and lifestyle as well as food security. Inextricably linked to environmental damage is damage to Indigenous peoples' cultural heritage and identity. The devastation of sacred sites, burial places and hunting and gathering spaces, not to mention a changing and eroding landscape, cause great distress to Indigenous peoples.

It would be a shame if options for conciliatory, interest-based negotiation and agreement making were narrowed or closed completely by emerging law and policy, leaving litigation as the only avenue for Indigenous peoples in Australia to take for recognition, assistance or redress in relation to the impacts of climate change.[41] Litigation carries significant risks including adverse costs, time and emotional wastage and long-term damage to relationships between parties. Every effort should be made in designing new legislation and policy to avoid Indigenous communities finding themselves with litigation as the only option to secure the rights and resources needed to respond to climate change and adapt to its impacts (see, for general discussion, Gerrard, 2008; *Native Village of Kivalina and City of Kivalina v ExxonMobil Corporation and others*, 2008).

Summary and conclusion

In the face of emerging climate change law and policy, the space currently afforded to Indigenous peoples in Australia, in terms of the protection and preservation of their legal rights and interests in country and culture, is vulnerable. Recent reports and initiatives of State, Territory and Federal governments signal an awareness of this issue and the potential opportunities climate change presents for Indigenous communities; however more than cognisance is needed to maximise participation and minimise additional burdens for Indigenous peoples.

---

40   And subsequent declarations and principles set out at UNFCCC Conferences since 2000, including media and reports of the UN Permanent Forum on Indigenous Issues: <http://www.un.org/esa/socdev/unpfii/> and referred to in this paper.

41   In Australia, to date, climate related legal action has focused on administrative action against governments and decision makers in planning and environment decisions, with varying degrees of success – see: *Australian Conversation Foundation & Ors v Minister for Planning* [2004] VCAT 2029 (29 October 2004), *Wildlife Preservation Society of Queensland Proserpine/Whitsunday Branch Inc v Minister for the Environment and Heritage* [2006] FCA 736, *Gray v Minister for Planning & Ors* [2006] NSWLC 720; *Queensland Conservation Council v Xstrata Coal Queensland Pty Ltd & Ors* [2007] QCA 338 and *Gippsland Coastal Board v South Gippsland SC & Ors (No 2) [2008] VCAT 1545*. In contrast, legal action elsewhere in the world has involved negligence and nuisance claims, as well as administrative legal action. Climate-related litigation is a reality, particularly in the United States where action has been taken against private companies, administrative decision and government agencies. (for example: *Cox v Nationwide Mutual Insurance Company No.* 1:05 CV 436 (S.D. Miss. filed. 20 Sept 2005; *Massachusetts, et al, Petitioners v Environmental Protection Agency et al* 549 US (2 April 2007)). While these cases involve laws specific to the United States, analogous arguments may be drawn in relation to environmental and other laws in Australia.

As Australia transitions to a carbon constrained future, which includes market mechanisms, there is a critical opportunity to address potential inequalities affecting Indigenous peoples. Establishing appropriate foundations now will provide effective pathways for development in the future. Without well-designed formal regulatory and institutional support the future of innovative carbon projects is likely to remain dependent on the goodwill of governments. Current reforms present a real and important opportunity to better value and support Indigenous peoples to exercise their right to development in accordance with their needs, interests and aspirations. In order to support a consistent and equal footing for Indigenous participation in emerging opportunities, it is important to further develop and more broadly apply measures similar to the initial positive steps taken in the CFI Bill. Further, it is critical to revisit policies and practices of conditional land return to Indigenous communities.

It is in the public interest for Indigenous peoples, as important knowledge holders and land managers, to be meaningfully and appropriately included in partnerships and in designing responses to climate change. Mitigation and adaptation strategies should provide the legal and practical infrastructure to facilitate the participation of valuable knowledge holders in formulating solutions to the issues facing our future livelihood.

Participation and empowerment, two basic and interrelated principles of the human rights-based approach to development, are particularly important for Indigenous peoples, who have been systematically excluded and marginalised from decision-making on matters affecting them (Stavenhagen, 2007). The involvement of Indigenous peoples in responses to climate change is not only consistent with the United Nations Declaration on the Rights of Indigenous Peoples and general human rights principles, as well as principles of national and international environmental law, it is also very likely to result in more comprehensive and appropriate solutions. The United Nations Declaration on the Rights of Indigenous Peoples provides an additional framework for development policies and actions that will affect Indigenous peoples.

Australia's response to climate change should incorporate the objects of existing domestic legislation or international instruments to which it is a party or has endorsed. A broad-reaching and complex problem necessitates an equally broad-reaching and diverse solution, in which the doors are open to future innovation, partnerships and economic opportunities for Indigenous peoples.

References

Aboriginal and Torres Strait Islander Social Justice Commissioner 2008, *Native Title Report 2007*, report to the Commonwealth Attorney General, Human Rights and Equal Opportunity Commission (now the Australian Human Rights Commission), Canberra.

The Albuquerque Declaration 1998, Proceedings from the 'Circles of Wisdom' Native Peoples/Native Homelands Climate Change Workshop-Summit, Albuquerque, New Mexico, 1 November 1998, available at: <http://www.nativevillage.org/Inspiration-/Albuquerque%20Convention.htm> (accessed 31 August 2011).

Allens Consulting 2005, *Climate Risk and Vulnerability: Promoting an Efficient Adaptation Response in Australia,* report prepared for the Australian Greenhouse Office, Department of Environment and Heritage, Canberra.

Altman, J 2001, *Sustainable Development Options on Aboriginal Land; The Hybrid Economy of the Twenty-first Century*, Discussion Paper No 226, Centre for Aboriginal Economic Policy Research, Australian National University, Canberra.

— and M Dillon 2004, *A Profit-related Investment Scheme for the Indigenous Estate,* Discussion Paper 270/2004, Centre for Aboriginal Economic Policy Research, Australian National University, Canberra, December.

— and K Jordan 2008, *Impact of Climate Change on Indigenous Australians: Submission to the Garnaut Climate Change Review*, CAEPR Topical Issue No 3/2008, Centre for Aboriginal Economic Policy Research, Australian National University, Canberra.

Anaya, J, J Henrikson and V Tauli-Corpuz 2009, *UN Experts Welcome Australia's Endorsement of the United Nations Declaration on the Rights of Indigenous Peoples,* United Nations Permanent Forum on Indigenous Issues, New York, available at: <http://www.un.org/esa/socdev/unpfii/documents/Australia_endorsement_UNDRIP.pdf>

Australian Human Rights Commission 2010, *The Community Guide to the UN Declaration on the Rights of Indigenous Peoples*, Australian Human Rights Commission, Sydney.

Australian Labor Party 2009, *Australia Supports Declaration on Indigenous Peoples*, media release, 3 April 2009, Canberra.

Banskota, B, S Karky and M Skutch (eds) 2007, *Reducing Carbon Emissions through Community-managed Forests in the Himalaya,* Kamal International Centre for Integrated Mountain Development, Kathmandu.

Bartlett, R 2004, *Native Title in Australia,* 2nd edn, LexisNexis Butterworths Australia, Chatswood.

Bernstein, L, P Bosch and O Canziani 2007, *IPCC Fourth Assessment Report: Climate Change 2007: Synthesis Report: Summary for Policymakers,* Intergovernmental Panel on Climate Change, Geneva.

Bird, DW, RB Bird and CH Parker 2005, 'Aboriginal burning regimes and hunting strategies in Australia's Western Desert', *Human Ecology* 33(4): 443–463.

Brundtland Report – see *Report of the World Commission on Environment and Development*

Bureau of Meteorology, *Indigenous Weather Knowledge Project,* available at: <http://www.bom.gov.au/iwk/ > (accessed 12 August 2011).

Campbell, D, J Davies and J Wakerman 2007, *Realising Economies in the Joint Supply of Health and Environmental Services in Aboriginal Central Australia,* Desert Knowledge CRC Working Paper 11, September 2007.

Cardinoza, M 2005, *Reviving Traditional NRM Regulations (Tara Bandu) as a Community-based Approach of Protecting Carbon Stocks and Securing Livelihoods,* Centre for International Forestry Research, Jakarta.

Castro Diaz, E 2008, 'Climate change, forest conservation and Indigenous peoples rights', Discussion paper prepared for the 'International Expert Group Meeting on Indigenous Peoples and Climate Change', Darwin, Australia, 2–4 April 2008, United Nations Permanent Forum on Indigenous Issues, New York.

CISDL Legal Brief 2002, 'The principle of common but differentiated responsibilities: origin and scope', Paper produced for the 'World Summit on Sustainable Development', Johannesburg, South Africa, 26 August–4 September 2002, United Nations, New York, available at: <http://www.cisdl. org/pdf/brief_common.pdf> (accessed 31 October 2008).

DCC – *see* Department of Climate Change

DCCEE – see Department of Climate Change and Energy Efficiency

Department of Climate Change 2005, *Planning Forest Sink Projects, A Guide to Legal, Taxation and Contractual Issues*, Australian Greenhouse Office, Australian Government Solicitor, Department of Climate Change, Canberra chapter 2.

— 2008a, *Australian Government Carbon Pollution Reduction Scheme Green Paper*, July 2008, Department of Climate Change, Canberra.

— 2008b, *Australian Government Carbon Pollution Reduction Scheme White Paper*, December 2008, Department of Climate Change, Canberra.

— 2008c, *National Carbon Offset Standard Discussion Paper*, December 2008, Department of Climate Change, Canberra.

— 2008d, *Draft National Carbon Offset Standard,* December 2008, Department of Climate Change, Canberra.

— 2009a, 'Indigenous Carbon Market engagement commitment', Document circulated and discussed at the Department of Climate Change (DCC).

— 2009b, *CPRS Exposure Draft Legislation,* Department of Climate Change, Canberra.

—, *Fact Sheet: Climate Change – Potential Impacts and Costs*, available at: <http://www.climatechange.gov.au/impacts/costs.html> (accessed 31 October 2008).

—, *Emissions Trading Timetable*, see: <http://www.climatechange.gov.au/emissionstrading/timetable.html> (accessed 31 October 2008).

Department of Climate Change and Energy Efficiency 2011a, Carbon Credits (Carbon Farming Initiative) Bill 2011.

— 2011b, 'Clean Energy Future, including Carbon Pricing Mechanism' – see <http://www.cleanenergyfuture.gov.au/> (accessed 12 August 2011).[42]

Department of Environment and Heritage (DEH) 2004, *Ways to Improve Community Engagement; Working with Indigenous Knowledge in Natural Resource Management*, Department of Environment and Heritage, Canberra.

— 2005, *Planning Forest Sink Projects – A Guide to Legal, Taxation and Contractual Issues*, Australian Greenhouse Office, Department of Environment and Heritage, Canberra.

---

42 Australian Government Carbon Pollution Reduction Scheme Green Paper/White Paper consultation process (July – December 2008) and exposure draft legislation (March 2009), CFI consultation and subsequent legislation, December 2010 and the recent Clean Energy Future 2011 package of reform. See: <http://www.climatechange.gov.au/cfi> (accessed 12 August 2011), and <http://www.cleanenergyfuture.gov.au/> (accessed 12 August 2011).

DEH – *see* Department of Environment and Heritage

Department of Environment Water Heritage and the Arts (DEWHA) 2009, 'Climate Change and Indigenous Peoples workshop', Canberra, Australian Capital Territory, 19 March 2009, Department of Climate Change, Canberra.

Department of Families, Housing, Community Services and Indigenous Affairs (FAHCSIA) 2009, *Closing the Gap on Indigenous Disadvantage: The Challenge for Australia,* Department of Families, Housing, Community Services and Indigenous Affairs, Canberra.

FAHCSIA – *see* Department of Families, Housing, Community Services and Indigenous Affairs

Department of Sustainability and Environment (DSE) 2008, *Victorian Government Land and Biodiversity at a Time of Climate Change: Green Paper,* Department of Sustainability and Environment, Melbourne, Victoria.

— 2009, Securing our Natural Future: A white paper for land and biodiversity at a time of climate change, Department of Sustainability and Environment, Melbourne, Victoria, <http://www.dse.vic.gov.au/conservation-and-environment/land-and-biodiversity-white-paper-securing-our-natural-future> (accessed 31 August 2011).

DSE – *see* Department of Sustainability and Environment

Department of Prime Minister and Cabinet (DPMC) 2001, *Climate Change Working Paper–Maori Issues,* New Zealand Climate Change Programme, Department of Prime Minister and Cabinet, Wellington, New Zealand, available at: <http://www.mfe.govt.nz/publications/climate/maori-oct01/maori-oct01.pdf> (accessed 31 August 2011).

DPMC – *see* Department of Prime Minister and Cabinet

Dodson, M 2007, *Report of the Secretariat on Indigenous Traditional Knowledge,* UN ESCOR, 6th session, UN Doc E/C.19/2007/10, United Nations, New York.

Garnaut, R 2008, *The Garnaut Climate Change Review,* Commonwealth of Australia, Cambridge University Press, Melbourne.

— 2011, *The Garnaut Review 2011. Australia in the Global Response to Climate Change.* Commonwealth of Australia, Cambridge University Press, Melbourne.

Gerritsen, R 2007, Constraining Indigenous Livelihoods and Adaptation to Climate Change in SE Arnhem Land, Australia, paper for the conference 'International Expert Group Meeting on Indigenous Peoples and Climate Change', 2007/WS.7, United Nations, New York.

Gerrard, E 2008, *Impacts and Opportunities of Climate Change: Indigenous Participation in Environmental Markets,* Australian Institute of Aboriginal and Torres Strait Islander Studies, Canberra.

Graham-Harrison, E 2007, 'Indigenous people fear double climate hit',*Reuters Media UK,* 13 December 2007.

Green, D 2006a, *How Might Climate Change Affect Island Culture in the Torres Strait,* Marine and Atmospheric Research Paper 011, Commonwealth Scientific and Industrial Research Organisation, Australia.

— 2006b, *Climate Change and Health: Impacts on Remote Indigenous Communities in Northern Australia,* Marine and Atmospheric Research Paper 012, Commonwealth Scientific and Industrial Research Organisation, Australia.

Heckbert, S, J Davies, J Cook, J McIvor, G Bastin and A Liedloff 2008, *Land Management for Emissions Offsets on Indigenous Lands,* CSIRO Sustainable Ecosystems, Townsville, Alice Springs, Darwin, Brisbane.

Hepburn, S 2008, 'Carbon rights as new property: towards a uniform framework', Seminar paper presented at 'ANU College of Law Seminar Series', Canberra, 21 August 2008, Australian National University, Canberra.

Insley, CK and R Meade 2008, *Māori Impacts from the Emissions Trading Scheme: Detailed Analysis and Conclusions,* 37 Degrees South Limited and Cognitus Advisory Services Limited, prepared for the New Zealand Government, Ministry for the Environment, Wellington.

Intergovernmental Panel on Climate Change (IPCC) 2004, *16 Years of Scientific Assessment in Support of the Climate Convention,* Intergovernmental Panel on Climate Change, Geneva, available at: <http://www.ipcc.ch/pdf/10th-anniversary/anniversary-brochure.pdf> (accessed 31 August 2011).

— 2007, *WGII Fourth Assessment Report, Summary for Policy Makers,* Intergovernmental Panel on Climate Change, Geneva, available at: <http://www.ipcc.ch/> (accessed 31 August 2011).

IPCC – *see* Intergovernmental Panel on Climate Change

International Council on Human Rights Policy (ICHRP) 2008, *Climate Change and Human Rights, a Rough Guide,* International Council on Human Rights Policy, Versoix, Switzerland, available at: <www.ichrp.org/files/reports/36/136_report.pdf> (access date unknown).

International Forum of Indigenous Peoples and Climate Change (IFIPCC) 2007, 'Reduced emissions from deforestation and forest degradation (REDD)', Statement made at conference '13th session United Nations Framework

Convention on Climate Change (UNFCCC) Conference of the Parties', November 2007, available at: <http://archive.forestpeoples.org/documents/ forest_issues/unfccc_bali_ifipcc_statement_redd_nov07_eng.shtml> (accessed 31 August 2011).

IFIPCC – *see* International Forum of Indigenous Peoples and Climate Change

International Work Group for Indigenous Affairs 2008, 'Conference on Indigenous Peoples and Climate Change: Meeting Report', Paper submitted to the 'Permanent Forum on Indigenous Issues, seventh session', New York, 21 April–2 May 2008, E/C.19/2008/CRP.3, United Nations, New York.

The Kyoto Protocol to the United Nations Framework Convention on Climate Change 1998, United Nations Framework Convention on Climate Change, United Nations, New York, available at: <http://unfccc.int/resource/docs/ convkp/kpeng.pdf> (accessed 31 August 2011).

Langton, M, M Tehan, L Palmer and K Shain (eds) 2004, *Honour Among Nations? Treaties and Agreements with Indigenous People,* Melbourne University Press, Melbourne.

Langton, M, O Mazel and L Palmer 2006a, 'The "Spirit" of the Thing: the boundaries of Aboriginal economic relations at Australian Common Law', *The Australian Journal of Anthropology* 17(3): 307–321.

Langton, M, O Mazel, L Palmer, K Shain and M Tehan (eds) 2006b, *Settling with Indigenous People,* Federation Press, Sydney.

Law, R, J Davies and J Childs 2007, *Examples of NRM Contracting within Australia*, Working Paper 7, Desert Knowledge CRC, Alice Springs.

Lyster, R and A Bradbrook 2006, *Energy Law and the Environment*, Cambridge University Press, Melbourne.

Macchi, M, G Oviedo, S Gotheil, K Cross, A Boedhihartono, C Wolfangel and M Howell 2008, *Indigenous and Traditional Peoples and Climate Change,* Issues Paper, International Union for the Conservation of Nature, available at: <http://cmsdata.iucn.org/downloads/indigenous_peoples_climate_change. pdf> (accessed 31 August 2011).

Macklin, J and W Snowdon 2007, *Indigenous Economic Development, Rudd Labor Government Election 2007 Policy*, Australian Labor Party, Canberra.

The Marrakech Accords 2001, Seventh Conference of the Parties to the UNFCCC (COP 7), Marrakech, Morocco, 29 October–19 November 2001, United Nations, New York.

Murdiyarso, D and H Herawati (eds) 2005, *Carbon Forestry – Who will Benefit?: Proceedings from Workshops on Carbon Sequestration and Sustainable Livelihoods*, Center for International Forestry Research, Jakarta.

Murray Lower Darling Rivers Indigenous Nations (MLDRIN), available at: <http://www.mldrin.org.au/about/default.htm>

MLDRIN – *see* Murray Lower Darling Rivers Indigenous Nations.

Northern Territory of Australia Garnaut Review Case Study 2008, *Abating Greenhouse Gas Emissions through Strategic Management of Savanna Fires*, available at: <http://www.garnautreview.org.au/2008-review.html/> (accessed 31 August 2011).

Peel, J 2007, 'The role of climate change litigation in Australia's response to global warming', *Environment and Planning Law Journal* 24(2): 90–105.

Report of the United Nations Conference on Environment and Development, Rio de Janeiro, 3–14 June 1992, 1992, A/CONF.151/26 (Vol I), United Nations, New York.

Report of the World Commission on Environment and Development: Our Common Future (the Brundtland Report) 1987, United Nations, New York, available at: <http://www.un-documents.net/wced-ocf.htm> (accessed 31 August 2011).

Russell-Smith, J 2007, 'Emissions abatement opportunities from Savanna burning', Paper presented at 'Workshop for Greenhouse Emissions Offsets Programs', Melbourne, July 2007.

Salick, J and A Byg (eds) 2007, *Indigenous Peoples and Climate Change*, Tyndall Centre for Climate Change Research, University of Oxford, Oxford.

Sands, P 2003, *Principles of International Environmental Law*, Second Edition, Cambridge University Press, United Kingdom.

Secretariat of the United Nations Permanent Forum on Indigenous Issues 2007, *Climate Change – An Overview*, United Nations Department of Economic and Social Affairs, United Nations, New York.

Smith, J and AJ Scherr 2002, *Forest Carbon and Local Livelihoods: Assessment of Opportunities and Policy Recommendations*, CIFOR Occasional Paper No 37, Centre for International Forestry Research, Jakarta.

Solomon, S, D Qin, M Manning, Z Chen, M Marquis, KB Averyt, M Tignor and HL Miller (eds) 2007, *Climate Change 2007: The Physical Science Basis. Contribution of Working Group I to the Fourth Assessment Report to the*

Intergovernmental Panel on Climate Change Cambridge University Press, Cambridge, United Kingdom and New York, United States, available at: <http://www.ipcc.ch/> (accessed 31 August 2011).

Stavenhagen, R 2007, *Promotion and Protection of all Human Rights, Civil, Political, Economic, Social and Cultural Rights, Including the Rights to Development,* Report of the UN Special Rapporteur on the situation of human rights and fundamental freedoms of Indigenous peoples, A/HRC/6/15, 15 November 2007, United Nations, New York.

Strelein, L 2006, *Compromised Jurisprudence, Native Title Cases since Mabo,* Aboriginal Studies Press, Canberra.

Tauli-Corpuz, V, and A Lynge 2008, 'Impact of Climate Change Mitigation Measures on Indigenous Peoples and on their Territories and Lands', Report for the 'United Nations Permanent Forum on Indigenous Issues, Seventh Session', New York, 20 March 2008, E/C.19/2008/10, United Nations, New York.

Tauli-Corpuz, V and P Tamang 2007, 'Oil Palm and Other Commercial Tree Plantations, Monocropping: Impacts on Indigenous Peoples' Land Tenure and Resource Management Systems and Livelihoods', Paper presented at 'United Nations Permanent Forum on Indigenous Issues, Sixth Session', New York, 14–25 May 2007, E/C.19/2007/CRP.6. United Nations, New York.

Tropical Savannas Cooperative Research Centre, *The West Arnhem Land Fire Abatement Project (WALFA)*, available at: <http://www.savanna.org.au/savanna_web/information/arnhem_fire_project.html> (accessed 31 August 2011).

United Nations Development Programme (UNDP) 2007, *Fighting Climate Change: Human Solidarity in a Divided World*, Human Development Report 2007/2008, United Nations Development Programme (UNDP), available at: <http://hdr.undp.org/en/reports/global/hdr2007-2008/chapters/> (accessed 31 August 2011).

UNDP – *see* United Nations Development Programme

United Nations Framework Convention on Climate Change (UNFCCC) 1994, United Nations, New York, available at: <http://unfccc.int/essential_background/convention/items/2627.php> (accessed 31 August 2011).

UNFCCC – *see* United Nations Framework Convention on Climate Change

United Nations Permanent Forum on Indigenous Issues (UNPFII) 2008, Meeting Report E/C.19/2008/CRP.3, 'Conference on Indigenous Peoples and Climate

Change, seventh session', Copenhagen, Denmark, 21–22 February 2008, United Nations Permanent Forum on Indigenous Issues, United Nations, New York.

UNPFII – *see* United Nations Permanent Forum on Indigenous Issues

Von Doussa, J, A Corkery and R Charters 2008, *Human Rights and Climate Change; Background Paper,* Human Rights and Equal Opportunity Commission, Sydney.

Wong, P 2009, '$10 Million to help Northern Australian Indigenous Communities Explore Potential of Future Carbon Markets', media release, 20 March 2009, available at: <http://www.environment.gov.au/minister/archive/env/2009/mr20090320a.html> (accessed 31 August 2011).

Woodward J 1974, *Aboriginal Land Rights Commission, Second Report*, AGPS, Canberra.

Working Group I of the Intergovernmental Panel on Climate Change (IPCC), 2007, *The Physical Science Basis*, Working Group I Report, The Fourth Assessment Report of the Intergovernmental Panel on Climate Change (IPCC), available at: <http://www.ipcc.ch/publications_and_data/ar4/wg1/en/contents.html> (accessed 31 August 2011).

Working Group II of the Intergovernmental Panel on Climate Change (IPCC), 2007, *Impacts, Adaptation and Vulnerability*, Working Group II Report, The Fourth Assessment Report of the Intergovernmental Panel on Climate Change (IPCC), available at: <http://www.ipcc.ch/publications_and_data/ar4/wg2/en/contents.html> (accessed 31 August 2011).

Working Group III of the Intergovernmental Panel on Climate Change (IPCC), 2007, *Mitigation of Climate Change,* Working Group II Report, The Fourth Assessment Report of the Intergovernmental Panel on Climate Change (IPCC), available at: <http://www.ipcc.ch/publications_and_data/ar4/wg3/en/contents.html> (accessed 31 August 2011).

Wotjobaluk, Jaadwa, Jadawadjali, Wergaia and Jupagalk native title determinations: what they mean for the Wimmera region, 2005. National Native Title Tribunal, Perth, available at: <http://www.nntt.gov.au/News-and-Communications/Media-Releases/Documents/2005%20media%20release%20attachments/2005%20Wimmera_Wotjobaluk%20deetrmination%20background%20and%20map.pdf> (accessed 31 August 2011).

Cases

Akiba on behalf of the Torres Strait Islanders of the Regional Seas Claim Group v Queensland (No 2) [2010] FCA 643 (*"Akiba"*)

Commonwealth v Yarmirr [1999] FCA 1668

Fejo v Northern Territory [1998] HCA 58; 195 CLR 96

James on behalf of the Martu People v State of Western Australia [2002] FCA 1208

Kaurareg People v Queensland [2001] FCA 657

Mabo v Queensland [No. 2] (1992) 175 CLR 1

Mabuiag People v Queensland [2000] FCA 1065

Masig People v Queensland [2000] FCA 1067

Members of the Yorta Yorta Aboriginal Community v Victoria [1998] FCA 1606

Members of the Yorta Yorta Aboriginal Community v Victoria [2002] HCA 58

Native Village of Kivalina and City of Kivalina v ExxonMobil Corporation and others, 2008, Complaint for Damages and Demand for Jury Trial, (US District Court, Northern District of California, 28 USC ßß 1331, 2201)

Northern Territory v Alyawarr, Kaytetye, Warumungu, Wakaya Native Title Claim Group (2005) 145 FCR 422 at [157]

Neowarra v State of Western Australia (2003) 2003] FCA 1402

Saibai People v Queensland [1999] FCA 158

Western Australia v Ward [2002] HCA 28

Wik Peoples v Queensland (1996) 187 CLR 1

Yanner v Eaton (1999) 201 CLR 351

Legislation

Aboriginal Land (Manatunga Land) Act 1992 (Vic)

Aboriginal Lands Act 1991 (Vic)

Charter of Human Rights and Responsibilities Act 2006 (Vic)

Environment Protection and Biodiversity Conservation Act 1999 (Cth)

Human Rights Act 2004 (ACT)

Native Title Act 1993 (Cth)

Racial Discrimination Act 1975 (Cth)

Covenants and Conventions

International Covenant on Civil and Political Rights, 16 December 1966, 999 UNTS 171

International Covenant on Economic, Social and Cultural Rights, 16 December 1966, 993 UNTS 3

International Labour Organisation Convention (No. 169) Concerning Indigenous and Tribal Peoples in Independent Countries, Arts 15, 23

United Nations Convention on Biological Diversity (UNCBD), 1992. United Nations, New York. Available from: <http://www.cbd.int/doc/legal/cbd-un-en.pdf> (access date unknown)

Declarations

Declaration of Indigenous Peoples on Climate Change, 2000, Delivered at the second 'International Indigenous Forum on Climate Change', The Hague, The Netherlands, 15 November 2000, available at: <http://www.austlii.edu.au/au/journals/AILR/2002/18.html> (accessed August 2011).

Declaration of the First International Indigenous Forum on Climate Change (the Lyon Declaration), 2000, available at: <http://www.treatycouncil.org/new_page_5211.htm> (access date unknown).

Rio Declaration on Environment and Development, Principle 22, UN Doc A/CONF.151/26 (Vol 1) (1992).

United Nations Declaration on the Rights of Indigenous Peoples (UNDRIP), 2007. United Nations, New York.

Universal Declaration of Human Rights, GA Res 217A (III), UN Doc A/810.

www.ingramcontent.com/pod-product-compliance
Lightning Source LLC
Chambersburg PA
CBHW061222270326
41927CB00022B/3452